Creative Thinking & Problem Solving

John Fabian

LEWIS PUBLISHERS

Library of Congress Cataloging-in-Publication Data

Fabian, John, 1934–
    Creative thinking and problem solving / John Fabian.
      p. cm.
    Includes bibliographical references.
    1. Creative ability in science. 2. Problem solving. I. Title.
Q172.5.C74F33   1990
153.4'3—dc20   90–31911
ISBN 0–87371–153–X

COPYRIGHT © 1990 by LEWIS PUBLISHERS, INC.
ALL RIGHTS RESERVED

Neither this book nor any part may be reproduced or transmitted in any form or by any means, electronic or mechanical, including photocopying, microfilming, and recording, or by any information storage and retrieval system, without permission in writing from the publisher.

LEWIS PUBLISHERS, INC.
121 South Main Street, Chelsea, Michigan 48118

PRINTED IN THE UNITED STATES OF AMERICA

This book is dedicated to:

My folks, who whetted my appetite for creative expression

Mentors galore who showed me the imaginative potency and wisdom of the human mind, teams at work and play, organizations seeking innovation and transformation.

Detractors whose probes and barbs helped hone my thinking and teaching.

Peers, students, and clients who cultivated a curious air, a playful air, an always-ready creative dare.

Pioneers who tinkered and toiled, tantalizing me with their inventions.

John P. Fabian, PhD, is a widely experienced and highly effective consultant, trainer, lecturer, and facilitator. For many years he has been helping diverse companies, institutions, groups, and individuals creatively untangle problems and opportunities, find new options, explore the future, and behave more openly and imaginatively. Through the use of creative thinking strategies, he has helped scientists, engineers, researchers, managers, and support staff discover novel approaches in their work, and ways of unlocking their own creativity and potential.

A licensed psychologist, Dr. Fabian helps bridge walls between organizations, disciplines, teams, and individuals. Thinking strategically and creatively, building high performance teams and organizations, solving thorny problems, and exciting people with the potential within and around them are the targets of his work.

Preface

Writing a book like this is an audacious act! Not wanting a tedious textbook nor a fluffy generic self-help book, I've attempted a balancing act. To provide the meat of creative thinking with the right amount of tenderizer, I've tried to present the important concepts and techniques without heavy gristle of background information that could detract from the taste and usefulness of the ideas.

The pages ahead are intended for scientists, engineers, and project leaders who want to add depth, how to's, and spice to their creative thinking. This book is for those who want to sharpen their imaginative edge:

- Scientists probing the boundaries of the known or searching for answers to perplexing problems or exciting possibilities in the field or the lab
- Engineers at the forefront of product or process development, as well as those maintaining, patching, enhancing, or seeking solutions to technology that already exists
- Project managers wanting to forge teams not only driving toward milestones but also applying imagination to their tasks
- Others tantalized by creativity

This book doesn't guarantee a Nobel prize, a seat on a prestigious commission of one's choosing, a huge grant from a foundation, or a new product that takes the nation by storm. Many factors besides innovative thinking go into those prizes.

Other opportunities lie in store for you in this book:

- Practical techniques for sparking individual and group creative thinking
- Expanded inventive behaviors
- Ways for people with different thinking preferences to interact creatively

- Methods of breaking common perceptions and producing wider ranges of options or ideas
- Remedies for overcoming innovation killers
- Greater appreciation for where, when, and how creative thinking can be useful in your work

If you already know a subject, you might want to move to another section. You may then want to double back for either concepts or how to's. The following thumbnail sketch should help you know what to anticipate in the pages ahead.

- *Chapter 1:* Sketches the creative thinking journey that has been and can be taken by many scientific and technical people. Provides definitions. Presents an overview of critical components.
- *Chapters 2 and 3:* Point to elements that help unleash creative energy. Highlight the individual qualities that deliver breakthrough thinking. Show blocks to creative thinking and ways to get around the personal barriers.
- *Chapters 4–9:* Explore how to ignite imagination. Go step-by-step through the creative process, illustrating the four key phases.
- *Chapter 10:* Provides specific practice and application tools for expanding imaginative thinking in individuals and groups.
- *Chapter 11:* Focuses on environmental keys for fostering and managing innovative thought.

The book presents a buffet of ideas, a mixture of individual and group stimulation concoctions. Fundamentals are given along with the fancier morsels. Just as professional teams prepare for sporting events by working on the basics, the foundation innovative thinking skills and methods need to be well in hand before more esoteric practices can jell.

You already have built-in imaginative abilities and creative techniques. I hope the ideas in this book can tease, stretch, and stir what you already have, then take you to richer levels of creative thinking.

Become a participant in the book. Keep your mind loose as

you search for applications. Have fun. Play with the concepts, the how to's, and the strategies. See what fits for you and what might work for your work group. Get involved when asked to.

The methods in this book come from a variety of sources. Many streams of thought come together here. Researchers and practitioners have been trying to ferret out the ingredients of creative thinking for many years. In the 1950s, divergent reasoning was being contrasted with convergent thought. During the 1960s, brainstorming and other creative problem solving approaches were being tested on college campuses and in many business settings. The human potential movement probed the imaginative mind and spirit as a principle of personal growth.

The 1970s was a time of explosive study of the contrasting modes of the two brain hemispheres. Neuroscientists, psychologists, and educators saw the historically mute right brain as the site of functions that paralleled elements believed to be critical to innovative thinking. The exploration of consciousness and its many states was another stream that added further substance and conjecture regarding the creative mind.

Also during this time, organization development specialists, in their attempts to revitalize life within companies, industries, and agencies, began to tackle work environments that inhibited creative thought. They began to create methods to breathe more imagination and innovation into organizations. Another important tributary for the study of creativity was the wealth of knowledge accumulating around the role of stress and change in the human mind and body.

During the 1980s and on into the 1990s, there has been much refinement and strengthening of earlier themes and areas of study. Creative reasoning has begun to be more balanced. It is seen as an integration of styles of thinking, emotions, levels of consciousness, and intuitive processes. Arational or whole mind thinking has been a target for stimulating creative thought. Assessments of preferred reasoning modes of individuals are expanding. The environment and how it affects both intrinsic and extrinsic motivation has been an intriguing focus. A wide variety of methods has been developed to elicit creative thinking.

My fascination with creativity and the use of imagination spans twenty-five years. I have applied creative thinking processes to a wide variety of engineering and scientific populations as both an internal and external organizational consultant and as an instructor for public and in-house workshops around the nation. In 1980, while at Battelle Memorial Institute's Pacific Northwest Laboratories, I studied the effect of creative thinking techniques on scientists, engineers, and other researchers. The result was a dissertation on creative stimulation and cognitive style.

From the 1960s to now, I have had many mentors in my own exploration of the creative mind, including colleagues, clients, academicians, and workshop participants. One who especially stands out is Bob Samples, an author, educator, researcher, philosopher, and seminal thinker, who stimulated my thinking and who provided instructional expertise during my dissertation study.

In my churning to produce this book, many people have been very kind in reviewing, suggesting, and supporting me at various stages. Robert Bailey, author of *Disciplined Creativity for Engineers* and former engineering professor at the University of Florida, made numerous excellent comments on my first draft and has provided me with fascinating written and phoned commentary. Harry Babad, a high-ranking scientist in the Pacific Northwest with an avocation for systems engineering and philosophy of science, has given me much personal time previewing two drafts. He has helped me define illustrations and clarify my writing. Roy Meador, a free-lance writer of scientific and technical topics, also made useful comments on the first draft.

Ken Gasper, a new product and technology director, Richard Trantow, an ultrasonics engineer, and Nancy Jones-Trantow, an organizational development consultant, gave me both technical and creative assistance for parts of my book. The person who has most supported and cajoled me, edited countless pages, and endured my giving-birth process is my wife, Ruth. I greatly appreciate their contributions.

John Fabian

Contents

Preface   vii

1   The Quest: Pursuing Creativity and Innovation

Introduction   1
    Getting inspiration from the past   3
    Experiencing *eureka*   5
    Starting the journey   6
    Taking giant or small steps   10
Creative thinking   12
    Debunking myths   12
    Defining key words   15
    Deciding who should be involved   17
    Choosing outcomes   19
Breakthrough creative elements   21
    The human instrument   23
    Discovery processes   23
    Enhancements   24
    Environments   24

2   The Human Instrument: Knowing Your Mental Equipment

Consciousness: focusing equipment   27
    Inner world   29
    Outer world   29
    Planetary world   30
The brain and mind: information processing equipment   31
    Psychological types   32
    Brain-mind modes   37
    Rational and arational modes   39
Processing and expressing creative thought   41
    Primary arational processes   42
    Media for expression   46
Summary   52

3 The Human Instrument: Expressing Breakthrough Qualities

Breakthrough quality 1: soft focus 55
    Hard, narrow attention 55
    Soft, open attention 56
Breakthrough quality 2: childlikeness 60
    Creative child within 60
    Beginner's mind 62
    Playfulness 63
    Exploration 64
    Fantasy 65
    Emotional coloring 66
Postscript 67

4 Breakthrough Discovery Process

The four phases 72
Looking ahead 75

5 The Target: Taking Aim

Situation–target interaction 77
Zooming in or out of the situation and target 80
Targeting steps 83
Case of the parched land 91
    Setting 91
    Concern 91
    Analysis 91
    Criteria/boundary conditions 93
    Target 93
    Solution availability 93
    Question prompt 94
Analysis methods 94
    Situation analysis 96
    Force field analysis 99
    Cause/consequence analysis 102
    Pareto principle 104
    Historical timeline 108

Observation   108
Value analysis   109
Arational mode postscript   111
Communication tips postscript   112

6   Search Keys: Igniting the Imagination

Mind-set breaking strategies   118
    Storm   119
    Calm   120
    Access   120
Search keys   121
    Spark a storm   122
    Follow storming rules   123
    Tinker with the angles   125
    Stimulate and relax the senses   125
    Use analogies   126
    Change your form   127
    Force fit   128
    Take a leap   128
    Extract   129
    Structure   129
    Access intuition and the preconscious   133
    Incubate   134
The beginning   136

7   Search: Following Steps and Strategy Formats

Steps in the hunt   139
    Decide outcome desired   139
    Choose, apply strategy format to match desired outcome   140
Strategy formats   142
    Brainstorming: the bread-and-butter process   142
    Brainwriting: verbal storms without group discussion   145
    Mindmapping   149
    Analogy storm   153
    Picture tour   161
    Imaging the future   165

Communication tips postscript 171
Shifting gears 174

8 Check: Assessing Options

Pull out the measuring tape 176
    Clarify ideas, if necessary 180
    Sort into themes and cull, if necessary 181
    Weigh, choose option(s) 181
    Search for more ideas 182
    A second look: analyze potential problems 182
Evaluation methods 183
    Clarify 183
    Sort and cull 184
    Weigh 188
Pause for perspective 199
Communication tips postscript 202

9 Action: Launching Ideas

Steps to make the idea real 206
    Develop plans to enhance option(s) and reduce potential problems 207
    Implement the proposals, tests, models, or simulations 211
Conviction and caution 212
Communication tips postscript 214

10 Enhancements: Enriching Individuals and Groups

The individual's grab bag 223
Tap group genius 234
    When to call on group genius 236
    Group enhancements: enriching group genius facilitation 237
Enhancement epilogue 243

11 Breakthrough Environments: Setting the Climate

Environments bursting with creative energy 248
    Build trust and freedom 250

    Put mouth and actions together   254
    Model creative behavior   257
    Build a porous, bendable organization   259
    Stimulate mind and senses   264
Environmental epilogue   270

Notes   271

Further Reading   277

References   285

Index   293

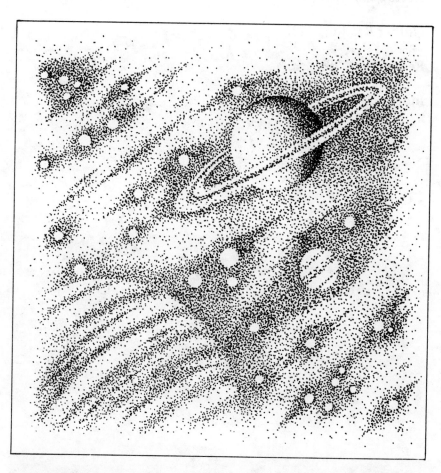

Within our minds are many worlds. As exciting as a journey into outer space, an exploration of our inner universe can yield the treasures and energy of the imagination. Such a trip beckons like a good adventure story. It's engrossing and consuming!

1 The Quest: Pursuing Creativity and Innovation

Chance favors the prepared mind.
—LOUIS PASTEUR

Most new discoveries are suddenly-seen things that were always there. A new idea is a light that illuminates presences which simply had no form for us before the light fell on them.
—SUSAN LANGER

INTRODUCTION

Are you a modern-day Ponce de León, wanting to make a great discovery? Have you stumbled onto the fountain of creative thinking? There is a wellspring of creativity and innovation that flows from the distant past to the modern present. Do your thoughts, tinkering, or theorizing sparkle with *aha*'s and creative energy? Then you are part of that stream of imagination.

Your quest for a momentous discovery could be taking place:

- On vacation, as you lazily think of new uses for miniaturization. Walking barefooted along the beach, you may conjure up mental images of miniature computerized probes tucked in the human body, automatically sensing hormone imbalances, circulatory pressures, hints of disease.
- In an out-of-the-way laboratory of your company, as you join your "tiger team" colleagues. You are engaged in a quiet but intense search for winning applications using a new gene-splitting process recently developed by your company.

- In a home shop or garage, as you try to improve a manufacturing process that has baffled your peers at work. You are intent on finding an answer, even though you have to step outside your discipline to do it.
- At your office, as you doodle with calculations to unlock a riddle for propulsion in outer space.
- In an organizational environment that tends to squelch imagination, as you search for new refrigerants that won't damage atmospheric ozone. You have to fight a steady chorus of "It can't be done" or "We don't have time to fool around with ideas that won't give us a quick payback."

Are you intrigued with the notion of tapping and enhancing creativity and producing more inventive ideas? You may already be brimming with imagination. If so, you may simply need a way to understand or to channel creative thinking more effectively—in yourself and in others. Or you may believe you came up short when creativity was passed around. If so, you likely need to catch a vision of your natural but partially hidden talent for inspired thinking. You most likely will also have to observe, then destroy, how you crush or muffle your creative thought.

Journeying into the land of creativity may cause you to be excited, yet also a bit wary. Creative thinking is an adventure of the mind. The venture requires alert, curious, exploratory thinking. The enterprise takes an apparent paradox—*disciplined* as well as *intuitive* reasoning. The undertaking needs skills that are seemingly inbred or learned, mental strategies that break old habits, techniques that can spark imagination in us or in groups. Although more schools today are trying to spawn inventive thinking, you may have been schooled to play down the use of imagination and play up rational and rote learning.

Many scientists and engineers are right at home balancing flexible, intuitive reasoning with logic and step-by-step processes. Others are uncomfortable with creative thinking because it can't be forced into a mold or a recipe. They become skeptical with processes that appear fuzzy when they first encounter them. As you move through this book, note your comfort level with the different ideas put forth. Give yourself permission to nibble or gorge yourself on the concepts, exercises, suggestions, and techniques.

Before you explore a creative thinking process that can effectively mobilize your imagination, take a little time in this chapter to get ready for the quest. You will look at what you need to consider in launching a creative effort—the seeds of inspiration and *eureka*, when to take the inventive journey, what imaginative thinking is and isn't, who to have join you, what you want to have happen, and how you get breakthrough discoveries.

Getting Inspiration from the Past

A glance back over our shoulders can provide inspiration, if we need it, for freeing our own creativity. Creative thoughts have been with humankind since the beginning. Scientists and engineers were plying their disciplines with imagination long before their fields were formalized and recognized. The wizardry of the mind has not been confined to any one era or location, although some periods of history and some cultures have produced more inspired and lasting products of imaginative thinking than others.

Majestic pyramids, simple but effective mining apparatus, fine paper, wood- and metal-cutting lathes, destructive gunpowder, and grand scale aqueducts are only a few products of ancient times. They attest to the disciplined and creative minds that left a legacy for their modern counterparts.

L. Sprague de Camp's fascinating *The Ancient Engineers* is but one of the chronicles of engineering derring-do of the distant past. When the tales from that book are added to numerous other records, there is a trail of technical marvels and feats from ancient to modern times. Such stories inspire an itching to unleash the imagination.

Do you have anyone who helped fan your creative flame? As I look at my own past, I have been inspired by mentors, colleagues, parents, and accounts of exploits of remarkable accomplishments over the distant and recent past.

My mother, who has her own creative spark, inspired our family with tales of her inventor grandfather, Dan Lehman, who set up a medicine shop in Nappanee, Indiana. He developed good remedies for upset stomachs and burns, pain and coughs, and sore throat and muscular aches.

Perhaps you have someone or some event that helped you hunger to make a great discovery or contribution. There are numerous stories from the past that could lure us into our own creative ventures.

Albert Einstein fantasized about what it would be like to ride a beam of light and formed his famous theory of relativity.[1,2]

Thomas Edison, possibly the first systems engineer, pursued many practical needs in the world with intuition and imagination. His creation of the microphone, phonograph, motion pictures, and automatic telegraph would have produced enough acclaim to have satisfied many an inventor. But he also gave us the light bulb and a whole lighting system to go with it.[3]

An engineer from the eighteenth century, Wilhelm Mayback, became fascinated with a perfume atomizer. It wasn't the perfume that beguiled him. He linked the process of mixing liquid and air, inventing the carburetor.[4]

Since creative thinkers often break the perception of what has been thought possible, they have sometimes endured the scoffing of both the public and peers.

In the late 1700s, Oliver Evans asked the Pennsylvania legislature for funds to develop a steam-driven carriage, but he was told that such talk could cause him to be declared insane. Evans flew in the face of such criticism by developing a steam-powered amphibious dredge.[5]

Fifteen-year-old Philo Taylor Farnsworth proposed an idea for a camera that could change images into electrical impulses, despite many professional engineers believing television broadcasting impractical. His contributions helped usher in the world of television.[5] More recently, Edwin Land conjured up an idea of an instant camera, only to be told by many professionals in the field that it wouldn't work. Yet his Polaroid camera broke new paths for photography.

Is it just doggedness or conviction that keeps individuals in hot pursuit of an idea, even after ridicule, management resistance, meager pay, and dead ends? No, there is an adventurous side to creating, an excitement and allure that captures the spirit and gives one purpose. The imaginative act is like a birth taking place in the mind. It has been described as the epitome of challenge, the setting aside of one's own ego in the heat of creation.

Experiencing *Eureka*

Words of Archimedes seem to capture the moment and essence of creation.

His task was to determine if a crown given his ruler was real gold. He knew the weight of a given volume of gold. However, the crown was irregular and contained ornaments. The scholar fretted for days, for his old measures would not work. One day he took a bath. In a relaxed state, he observed how the water rose around him as he lowered his body into the tub. The solution flashed in his mind. The crown would displace an amount of water equal to its own volume. "Eureka!" he exclaimed. "I have found it."[6]

The *eureka* experience captures the spirit and excitement of many a creator and inventor.

Until the time of Edison, innovation was often linked with the ingenuity and labor of single individuals, pioneers who saw gaps and sought to fill them or who had nagging questions and found some way to answer them. At Menlo Park, Edison, the grandest of the old-time tinkerers and inventors, gathered around him a team of technically minded people to aid his search for solutions to numerous practical problems. Team players, team contributions, and team solutions became increasingly important as knowledge and the complexity of projects grew like a streaking meteor in the twentieth century.

Today organizations rely on the efforts of both individuals and work groups. Companies set their talent on a search for product lines that will whet the "I want to buy it" appetite of more people. Businesses crave breakthrough systems and technologies that can help maintain, gain, or reestablish a market foothold. Industries want innovations that can take them in whole new directions.

The business, industrial, and research world would like to hear more *eureka*'s. Attempts to use imaginative thinking have possibly never been as widespread as the current fever to instill organizations with the quest to be innovative, vital, and competitive. Books and consultants have been heralding the need to search for, manage, and get a passion for excellence. They espouse entrepreneurship, "skunk" or special project teams, and innovation philosophies and actions that can bring organizational effectiveness and prized new products.

Starting the Journey

Launching the journey toward *eureka* requires at least three things. You need a readiness to unleash your creative mind. You have to seize opportunities in any work setting. You will have to take a pathway that increases the chances of hitting imaginative pay dirt.

Mental Readiness

Walking the path toward discovery starts first with a keen mental approach. A lulled mind dulls the creative edge. Scientists and engineers have a skeleton of knowledge and methodology. Their musculature is the rigorous execution of their discipline. Ideally their heart and soul are unflagging, child-eyed curiosity and imagination.

Years of technological and scientific schooling, however, can blunt the heart and soul. There is so much knowledge to be learned and so much emphasis on analysis and precision that our creative thinking can remain underdeveloped and underused. In the working world some environments allow the imaginative spark to be kindled, while other job climates dampen the creative flame.

A variety of mental traits can let the creative spirit loose, rather than squelch it. The animal kingdom provides metaphors to parallel the mental approach you need. Engage the world with the curiosity and playfulness of a kitten with a ball of string. Continually scan your world seeking perspective, like an eagle able to soar high yet swoop down to grasp a tasty morsel. Retain full sensory awareness, like a lioness tuned to the environment when protecting her cubs. Pursue ideas with skill and imagination, like a fox stalking its prey.

Alertness to Opportunities

The second requirement of searching for *eureka* is to look around you in every work setting for chances to use your imagination. Discoveries are waiting to happen at any one of the general phases of activity in which engineers and scientists work. Figure 1 shows the range of arenas where innovations can be made. The primary spheres of influence and application, especially by those who look at "glamorous" positions in the field, have been thought to be in research, development, and design. Yet there are also rich opportunities for applying creative thought to manufacturing/operations and marketing. In any of the five application phases shown in Figure 1, you could be assigned or could take on a vast range of tasks that could benefit from creative thought.

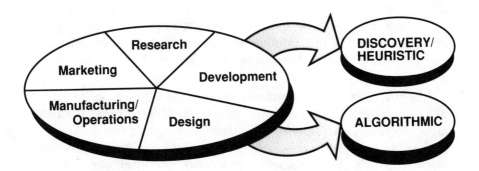

Figure 1. Broad application arenas and solution pathways.

Find a more efficient superconductor. Show the existence of quarks. Design a revolutionary new engine. Stop oil leaked into the ocean from damaging the ecological system of sea and air. Automate the movement of inventories to assembly lines. Develop a more powerful computer chip. The tasks are endless and come in varying sizes. These are just some of the engineering and scientific tasks that often require a fresh look, a touch of imagination, a novel approach, or a breakthrough thought. You may have to decide how much creative thinking is needed and how to apply it.

Pathways to Discovery

The third requirement in moving toward *eureka* is to take the path that gives you a good chance of making a breakthrough. When tackling a task in your work arena, you "draw a bead" on a puzzle or problem needing a solution. You compose hypotheses. You sketch scenarios or play "what if" games. You search for and test leads. When applying scientific principles to a practical problem or when trying to exploit a technological opportunity, you lay out the factors in the situation. You determine an approach. You hunt for ideas.

When someone hands you a problem to solve or you see an opportunity you'd like to explore, you decide what step to take next. To determine the right pathway for defining or solving a nagging, elusive problem, you may ask yourself a series of questions. Should you use a tried-and-true method of collecting and analyzing data? Is now the time to try a new way of doing the investigation? Should you go for commonplace solutions that might work, or decide to blow out the mental fuses and probe for a novel result?

Begin your creative journey when you don't have or don't like the existing algorithm for solving the problem or investigating an opportunity. An *algorithm is a tested, common approach—a chemical formula, a mathematical equation, a seasoned technique, a well-known measure.* Much of the discipline of science and engineering is the application of algorithms. It is a sound approach that makes use of a wealth of accumulated knowledge.

Sometimes the algorithm or solution doesn't fit or doesn't feel right. You have a hunch, a sense that more is needed. Some scientists and engineers would deny the validity of those peculiar feelings. They would bypass the wisdom of their body and intuition and try to make the old solution work. Unfortunately, they may miss the clue that could have led them to new possibilities.

If you are at a point of saying that fresh thoughts are needed, you are ready to begin the heuristic or discovery phase of your investigation. A *heuristic is an aid to creative problem solving. It is an experimental process or technique, a method for exploring and discovering solutions.* Describing useful heuristic or discovery processes, the tools of creativity, is the purpose of this book.

On your discovery path, you may apply algorithms along the way. But you will most likely encounter gaps, surprises, and uncharted territories. Pathways for finding a solution are many. Here are just a few routes toward discovery.

- **Trial and error:** In his search for the right material to burn in a glass bulb, Edison tediously tried out thousands of filaments. His engineering heart and soul persisted until he came across bamboo in a fan. Edison tore off a strip to play with, then tested it. Eureka! Charles Kettering uses the term *trial and success* to emphasize how this approach can be enhanced by keen observation and by learning from what didn't work.
- **Happy accident:** Michael Zasloff sat in his laboratory watching a frog swim in the muddy water of a tank. He had recently cut open and sutured the frog's stomach. In a contemplative mood, he started reflecting on why the frog wasn't getting infected. He had a hunch that it might be cecropins, a natural protective device found in some insects. He ground up the skin of the frog and tested it. Zasloff appears to have discovered a natural antibody.[7] This bit of serendipity required a mind that was relaxed, open, and curious.
- **A team approach,** applying a variety of creative thinking and problem solving techniques to get an imaginative result: Apple Computer, Inc. put together an inquiring, iconoclastic team sequestered from the main core of the organization. The team produced another *eureka*—a powerful, user-friendly personal computer that caught the imagi-

nation and allegiance of thousands of consumers. In his *The Soul of a New Machine,* Tracy Kidder captured a similar stirring story of a band of computer scientists and engineers in the discovery mode. Selective recruiting, long hours, an enigmatic but inspiring leader, and flashes of creative brilliance were some of the key factors that went into the creation of the Eagle, a new Data General computer.[8]

Once you have a mental readiness to apply your imagination to the tasks around you at work or home, you may choose from a variety of paths that can lead toward *eureka*. You could take the timeworn route to discovery, using a trial-and-error testing of solutions you think of by logic or by hunch. You could take the path of being vigilant and exploring, trying to bring about a serendipitous accident. You might pull a team together to tap group synergy. As you move along these pathways, however, you could improve your chances even further by applying heuristic or discovery processes that are sprinkled liberally throughout the rest of this book.

Taking Giant or Small Steps

How big does an assignment have to be before you can devote creative thinking to it? Some commentators say that it depends on whether you live in the United States or in another part of the world. There is a perceived difference, for instance, in the size of the tasks that are undertaken in the United States and in Japan.

Masaaki Imai makes such a contrast in *KAISEN: The Key to Japan's Competitive Success*, his book about the widespread Japanese process of making many minor improvements in businesses and industries. He sees the West as strong in innovation and weak in making improvements, while Japan is the opposite, weak in innovation and strong in making improvements.

A major effort in that country stresses quality. Workers get together in quality circles to tackle problems or enhancements in plants or on assembly lines. Statistical analysis is commonplace. Workers come up with numerous small, inch-like steps. At one company, for example, workers came up with many thousands of improvement steps in one year.

Imai believes that giant leaps of innovation are required at the research, development, and design activity phases listed in Figure 1. Smaller improvements are needed in the manufacturing/operations and marketing phases. Refinements and evolutionary changes prevent potentially good products from failing in the marketplace.[9]

During the research stage, bolder, more intuitive steps can often be taken. As one follows the phases from research to marketing, ideas become increasingly concrete and resistant to change. They are shaped, modeled, put in place, integrated into product lines and systems, and wrapped in fancy packages and marketing strategies. The winds of imagination need to blow through these later stages, as well as the early phases, to prevent industrial and business rigor mortis.

Until recently, American industry and business did neglect making substantial improvements in those latter phases. The United States' businesses have liked the home run, the breakthrough. A groundswell of attention is now being given to making improvements in the more practical arenas of production lines, heavy industry, service, and meeting needs of consumers. Some improvements will require giant steps, others will require enhancements by inches. Both are needed for new products and processes to become winners.

Japan, a consensus-laced society, is trying to gear up to more of the giant steps.[10] Recognizing their weakness in innovation, the Japanese set up a science city in the countryside around Kumamoto. This technopolis, the first in a series of such cities, is designed to be a center of technology and creativity.[11] They have also translated into Japanese Robert Bailey's *Disciplined Creativity for Engineers* to learn more about this somewhat culturally foreign subject.

In America, creative thinking cannot be reserved for the giant leaps. It must also be applied to improvements in small steps, gains by the inch and yard. John Gardner echoed this sentiment when he wrote in his *Self Renewal: The Individual and the Innovative Society:* "Many of the major changes in history have come about through successive small innovations, most of them anonymous."[12]

To accomplish the tasks large and small, tools and processes need to be applied. There is a point in the life of an assignment during which critical "how to proceed" decisions are made. Is an algorithm available or desirable? If it is, the pathway toward the solution will follow the logical application of the algorithm. If it isn't, a heuristic device or method needs to be applied.

The rest of the book strives to give you heuristic or creative thinking tools and strategies. Your analytical skills have already been developed by your formal education, so they won't be emphasized here. Instead we are concentrating on what it takes to enhance imaginative, intuitive, inventive types of reasoning.

CREATIVE THINKING

First I will debunk several common myths about creative thinking and define key words, then present guidelines about who to involve and what outcomes to target.

Debunking Myths

Folklore often surrounds popular notions of creative people and the thinking that produces imaginative ideas. We are awed by great creative acts or products. Mystique builds around creators—visionary researchers; pragmatic engineers; standout poets, artists, and musicians. To understand the kind of thinkers who make discoveries big and small and who apply imaginative reasoning, we have to challenge some of the myths that hover around this field. There are many myths, but we will explore just three critical ones.

Myth One: Creativity Requires Genius

Geniuses have made great, creative contributions. But genius does not a creator make. Creative products can be strokes of genius by ordinary folk. Correlations between intelligence and creativity haven't been strong. Although knowledge of a field is often critical for providing breakthroughs, genius may have little to do with the accumulation of knowledge.

You may be, like most of us, ordinary. You need not be blessed with the ability to do great feats of mental gymnastics. What you and the person down the hall have in common is the *potential* to be innovative.

Many inventors have just tapped that potential, applying plain old ingenuity to life's daily nuisances.

Floyd Paxton, for example, set out to find a way to close bags once they had been opened, a problem suggested to him by a man from Washington State's famous fruit country. Paxton whittled a piece of Plexiglas into a device now used around the world to close the plastic wrapping around bread.[13] It was an ordinary feat that became a packaging breakthrough.

You, too, can find ordinary or even unusual solutions to life's problems. You don't need a brain transplant. You already have the creative apparatus. The trick could be to get yourself or your mental straitjacket out of the way, to honor and use your intuitive and preconscious processes more effectively, or to use creative strategies during work or play. Belief in yourself and in the possibility of thinking more intuitively and creatively is a fundamental step.

*Myth Two: You Have to Be Odd
to Really Turn On the Imagination Afterburners*

Van Gogh didn't cut off his ear because he was creative. You can be innovative without cutting off your ear. The quality of being odd is not the essential ingredient of creative thinking. It isn't needed. The idea of the madcap inventor with wild hair, eccentric habits, and intense eyes tinkering with Rube Goldberg gadgets or mixing vials of strange concoctions in a laboratory dungeon is mainly spice for comics and science fiction.

Innovative thinkers can be single-minded, even stubborn in the pursuit of their ideas. They may challenge conventional viewpoints, probe beneath the surface, and look at life from unusual angles.

Depth of feeling rather than strange thinking, however, may be an ingredient of many creative people. Several recent studies have suggested that creative writers, for example, are overendowed with moodiness. Many have depressive or manic-depressive episodes. There are indications that volatility of mood may also be a characteristic of many creative people in other fields.[14,15,16]

To enhance your creativity, you don't have to work on accentuating the blues or swings of mood. Healthy allowance for feelings in your life and work, however, can loosen up the flow of imaginative power. Enjoying the emotions of wonder and creative excitement can stimulate inventive reasoning. Walking around with an emotional straitjacket is a handicap that doesn't allow your thoughts and your feelings to work together for more creative output.

Creative thinking is sound, not strange, reasoning. To create, you must be mentally and emotionally flexible yet persistent enough to extract and recombine threads from daily experience, your knowledge and image bank, the senses, and intuition. To be most effective, your mind and body should feel relaxed and open. Then your mental spinning can fashion new options and possibilities.

Odd people often don't think with clarity, freedom, and balance.

Myth Three: Creative Thinking Isn't Rigorous,
but Uses Simple Processes

Imaginative reasoning can be, like a good theory, elegantly simple. The problem comes in the execution of those processes or in the attitude or behavior that inhibits the release of creative thought. Imaginative reasoning can break down when people get sloppy, or when they don't apply a sense of discipline to their creative attack of a problem.

Individuals may say they use brainstorming to gather new ideas, for example, but instead only list ideas in the midst of a barrage of critical comments that actually restrain the flow of more novel thoughts.

Real creative thinking takes practice and involves carefully following principles and effective techniques. Those who want to be good at stimulating the imagination could take a lesson or two from athletes.

Professional athletes work hard to develop their craft. Out of the limelight they spend long hours rehearsing their specialty. A quarterback practices passing countless times, while receivers practice catching the ball over and over. Great runners will practice even off the track, creating a mental image of themselves in a race. They picture themselves crouching and poised on the track, bolting forward at the crack of the gun, sensing their breathing and pace stride for stride at each point in the race. Larry Bird, eminent basketball player for the Boston Celtics, commented during a national playoff series that he couldn't wait for the season to end so he could get back to practicing.

A disciplined approach hones the fundamentals—for development of the creative mind as well as for the dedicated athlete. It's practice, practice, practice that makes us consistently effective.

Defining Key Words

You have creative potential and don't have to be odd to release it, yet must apply discipline to develop it. What is this imaginative power that lets you dream up novel, practical, exotic, useful, wild, or out-of-the-ordinary thoughts? What is this thinking process that allows you to put on the hat of the explorer and start an adventurous search for golden ideas?

Many people have tried to define the creative process or mechanism. Definitions run the gamut from the practical to the philosophical to the technically-stated for research purposes. Here's a sampler:

"*Creativity is seeing what's really there.*"[17]

—SAMPSON RAPHAELSON

"*From one man, I learned that constructing a business organization could be a creative activity. From a young athlete, I learned that a perfect tackle could be approached in the same creative spirit.*"[18]
—ABRAHAM H. MASLOW

"*Creative thinking is really just good thinking.*"[19]

—GEORGE PRINCE

A definition fruitful for researching creativity:

"[Creativity is] the power of the imagination to break away from perceptual set so as to restructure anew ideas, thoughts, and feelings into novel and meaningful associative bonds."[20]
—JOSEPH KHATENA AND PAUL TORRENCE

When a researcher first read the next definition in one of my workshops, he thought the description was too mystical and philosophical. Having received permission to think more expansively, however, he went home and took a creative leap—coming up with a patentable idea. He said the words in this description did echo the state he was in while he was creating:

"When the inventive mode is functioning, a total synergic kind of knowing evolves. Exploration has the quality of a dream. Objects, processes, and conditions emerge, merge, and dissolve with no reason. The configurations that last do so because they have personal meaning, but not necessarily a public reason. Great courage is required at this stage because ego is transcended. Psychological fragilities are forgotten and they impose no retarding influence on the freedom of exploration. The total energies of intellectuality, emotionality, and sexuality merge into a force at once passive and aggressive. There is resolution to be, but not to control. The euphoria of discovery provides a continuing, peaceful orgasm of pan-existence. Giving and taking, dismissing and possessing all become the same. In this state of viable openness the invention, the creation, the peak experience ... happens."[21] —BOB SAMPLES

Other good, simple descriptions of creativity:

"[Imaginative reasoning is] an act that produces effective surprise."[22] —JEROME BRUNER

"The creative process is any thinking process which solves a problem in an original and useful way."[23] —H. HERBERT FOX

"... the totality of highly personal knowledge and art required to create."[24] —ROBERT BAILEY

Definitions for heuristic or discovery processes may seem fuzzy and imprecise, as hard to hold on to as soap in the shower. Creating isn't neat, clean, totally rational. It can be highly personal, disruptive, free-flowing, and work outside boundaries. Yet it's helpful to pin down a few definitions so we can have a common language. A number of intertwining terms have been used to describe various aspects of imaginative thinking. Here is a summary list:

- **Creative thinking:** *Mental processes that give birth to ingenious thoughts.* Reasoning that uses imagination to substitute, expand, modify, or transform the symbols, images, ideas, patterns, conditions, or elements in the world around us. Often contains an element of surprise.
- **Creativity:** 1. *Inventiveness.* 2. *Ability to create something new or imaginative.* May be measured by the newness to an individual, the infrequency with which people give a specific type of response, or an agreement that the product is both useful and original.
- **Innovation:** *Introduction of the results of creative thinking*—a new product, device, process, or method.
- **Invention (wears several hats):** 1. *That which has been created or discovered.* 2. *A creation that made it through the legal process and can wear the patent title.* 3. *Mental and behavioral processes of inventing.*

Heuristic thinking, then, is creative, inventive, and imaginative. Products of such thinking are innovations or inventions—or just plain good ideas. The words *creative, inventive,* and *imaginative* will be used interchangeably in this book.

Deciding Who Should Be Involved

When presented with a problem or opportunity, who do you include? Are you a Lone Ranger problem solver? If so, the choice is probably clear. You'll handle it yourself, applying whatever internal and external strategies you have applied many times in the past. Only occasionally will you involve a confederate or two.

Or you might like the team approach, a group of people to

engage in discussion and to arrive at a collective resolution. Involving others, of course, requires trust—the quality of being open, of sharing, of honoring someone else's ideas, and of receiving feedback on your thoughts and reactions.

Either the individual or the group approach could serve you well. Either way, you might bypass valuable contributors and input.

Consider what viewpoints should be considered to complement, challenge, or stretch your own. Begin by asking yourself a few questions:

- Assuming success, what will be the probable consequences? Who will be affected by any resolutions or actions?
- Who has the most expertise in this area? How can that knowledge and experience be tapped?
- Are organizational politics likely to be a major factor? How can we control that?
- Who has a successful track record creating new things?
- What parts of the organization—such as engineering, manufacturing, and marketing—have unique perspectives that should be considered? Who could be of help?
- Who will have to buy, sponsor, or evaluate the ideas that are chosen?
- Who can provide "fresh air" thinking or act as a devil's advocate when the situation is being examined, when creative thoughts are needed, or when evaluation and action are required?
- Are the skills of a facilitator needed to suggest creative processes, get a team unstuck, or prevent vested interests from blinding the problem-solving efforts?
- Are the perspectives of customers, stockholders, the opposition, or the community going to be represented? Who will be their agents?

When these questions are answered, more adequate consideration can be given to who should be involved and when. Combinations of individual, team, and organizational efforts are needed to adequately resolve many issues today.

Choosing Outcomes

People usually desire one or more of four general types of outcomes when they go after imaginative ideas. The types range from finding a better twist on something to producing the highly innovative, from gaining perspective on what's happening currently to stretching the mind with vision. Table 1 shows these desired outcomes in terms of increasing creativeness.

By having a good understanding of hoped-for results, you may tailor inventive thinking to match the situation. Choose an outcome and go for it with gusto. Once you are in an imaginative

Table 1. Range of Creative Outcomes

Increasing Degrees of Creativeness	Desired Outcomes	
	Better Twist	**Perspective**
	To change a process, mechanism, piece of equipment, or substance through adaptation or refinement. A breakthrough or novelty isn't needed.	To take a first-cut overview, to do a general diagnosis, or to see where to probe next.
	Innovation	**Vision**
	To create something different with methods, mechanisms, or materials, not just modify or sharpen what you have.	To develop a vision of the future or develop new strategies.
	To push the boundaries of your thinking, step outside your original definition, disregard perceived constraints, achieve new problem understanding and insights.	

mode, you may uncover new data that will change the outcome you desire. Just alter your expectations and, if needed, your methods.

If you use a team approach, for example, the outcome you want will help you shape your design of the creative effort. Do you want perspective or vision? Pick people who can think broadly and are willing to keep ideas approximate rather than nit-pickingly precise. Do you want an outcome with a better twist? Choose individuals who have hands-on experience in the area under consideration, people who can evaluate the practicality of the ideas as well as generate them. Do you want innovation? Include people who enjoy looking at the world from unique angles, can play with ideas, and don't mind allowing their thoughts to range into the zany and off-the-wall.

Robert Bailey, an inventive electrical engineer who worked in industry and who also taught creative thinking at the University of Florida for many years, has a formula for setting up small teams. The right team of three, he suggests, could outperform much larger teams. One person would be the synthetic thinker, a creative individual who adds ideas, thrust, and integration. A second member of the team is an analytic thinker, bailing the team out of subproblems with analysis. The third person is an expert in material resources, showing what kinds of materials to use where. Using such a team, perhaps creative results could range from better twists to innovation, perspective to vision.[25]

To get the outcome you need, you have a good start by knowing what level of creativity you want to target and getting a balanced mixture of people working with you, but don't kill yourself by shortchanging the time and the approach. Give yourself time to analyze the problem effectively by laying out and weighing the factors surrounding the situation.

If you want an innovative outcome, your team will need to clear their heads of commonplace ideas. Get away from jangling telephones and pestering colleagues. Go off-site. A two- or three-day retreat at a comfortable, informal hideaway may be very appropriate. Many executives almost faint when I suggest that. They cry: "Our managers have to be available for crises." "We have too many important things going on." "What would take all that time?" Yet sufficient quality time is needed to develop quality ideas. Commitment to innovation means using the time and the techniques that will increase the probability of getting more original payoffs.

BREAKTHROUGH CREATIVE ELEMENTS

Creating is a complex and exciting process. It entails the proper combination of numerous elements. Peak performance creative thinking combines the energy from each one of the elements. There is no magic in the components, but the result can have a magical quality to it. The outcome can be synergy, serendipity, and *eureka* for individuals or groups.

A technical team that has worked long and hard to corral some wild aspect of their project and ends in the wee hours with a highly imaginative outcome may feel exhilaration from their breakthrough idea. For that period of time, the energy from the creative components was strong enough to break old patterns of thought. The company is usually better off afterward, although at first it may have to be dragged kicking and screaming to implement the new creation.

A lone engineer or scientist may have spent days pouring over a research or production line problem. When the *aha* and excited *ah* arrived, the elements were synchronized, and their combined force was energizing.

The creative flow has two parts: elements and energy. Breakthrough *creative elements—the human instrument, discovery processes, individual and group enhancements, and environments—act as both the fuel and the engine for generating imagination.* They need to be vital, developed, well-honed. These four general elements or sources of energy form the basis for the rest of the book. Figure 2 displays the flow from the mixture of key ingredients to the desired results. Breakthrough creative elements are the keys to an imaginative outcome. When one of these elements is scrawny or damaged, it's as hard to create a spark of imagination as it is to run a car with sand in the engine. *Energy*, the second part of the creative flow, *is ignited by the breakthrough elements. Creative power is unleashed.* It disrupts mental habits and forces new perspectives. It

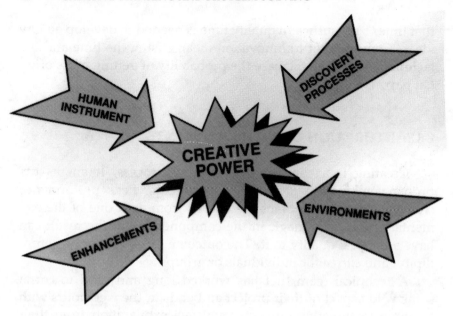

Figure 2. Creative elements—the sources of creative power.

turns intuition, dreams, and internal image making loose on knotty problems. It causes thoughts to tumble about and form new connections.

How do the breakthrough creative elements form the energy to force a breakthrough, or an advance in methods, knowledge, or products? When used together, they either energize or help release imaginative power.

A clue for their potency comes from a 1977 Nobel prizewinner in chemistry, Ilya Prigogine. The award was given for his idea that structures exchange energy with their environments. The more complex the structure—a chemical compound, a leaf, heavy traffic, people bunched together in a city, the brain—the more energy dissipates or is exchanged. The fluctuations of energy bring instability, often producing sudden shifts to new structures, new complexity, new patterns, new order. The interaction of energy brings both disequilibrium and innovation.[26]

Applied to creative thinking, Prigogine's dramatic ideas suggest that fluctuations in our approaches, thinking, and states of consciousness can trigger novel thoughts. Enhanced thinking and consciousness, energized thought processes, and enriched environments enable us to make the mental shifts to shake off our blinders and see new possibilities.

Here's a brief synopsis of the four elements that generate the imaginative power that leads to breakthroughs. The rest of the book enlarges upon each one of these components.

The Human Instrument

Of all the important tools at your disposal, the human instrument is critical. Like other human beings that inhabit this planet, you have amazing mental, perceptual, emotional, and motor equipment. When operating in top form, this human apparatus can become a formidable collection of personal power, producing breakthrough creative energy. Unless you are a rare species, you are likely aware of only a small portion of the creative potential encased in your body and mind.

Reasoning power sets you apart from birds, bees, bears, beetles, and baboons. Aside from degree of intelligence (which is not the telltale mark of creative thinking), how imaginatively you think is affected by your style of thinking, attitudes, traits, and emotional-perceptual characteristics. Your level of consciousness determines how open you are to the creative clues scattered around you. Your flow of creativity is helped or hindered by the quality of your "human equipment," but especially by how positively you approach yourself, your tasks, and your environment.

Discovery Processes

This breakthrough element contains two heuristic methods. The first is a process, the *creative* or *discovery cycle*. This is creative thinking broken into phases. It is a sequential process for solving problems and adds imaginative angles that can be used at each step. The creative cycle has been designed to capitalize on the wealth of innovative thinking research and practice. At each phase of the cycle are tips for making sure that, when using this discovery process in a group setting, interaction is open and vigorous.

The second key heuristic method for intentionally stimulating inventive reasoning uses *creative strategies*. This approach draws upon imagination-awakening techniques which are at the heart of the creative process. As blueprints, these arousal plans help you tackle problems and promising situations when a less rational and more imaginative approach is needed. These strategies can be simple, strange, complex, or exciting, and can be readily learned and cultivated over time.

Enhancements

These are individual and group tools, vehicles that can sharpen and expand your innovative thinking. Individual enhancements include a grab bag of techniques and exercises. Choose a few as starters. Practice with these methods can make your inner world more imaginative and vivid or stoke your mind with rich creative fuel. Group embellishments are designed to tap the genius of groups and eradicate what blinds teams. These suggestions can improve your skill in handling the dynamics of teams and in following an effective meeting format, thus releasing more synergy.

Environments

Numerous factors in the life around you can cause inventive reasoning to flourish or wither—encouraging attitudes from a manager, a crushing workload, inadequate technician support, jarring physical surroundings. Highly innovative, effective organizations succeed to the degree they create a culture in which innovation is esteemed, risk taking is prized, ideas are honored, communication flows openly, trust grows, and intrinsic motivation is cultivated.

Encouragement, support, nourishment, and reinforcement are key aspects of the ongoing efforts in companies to pursue a crusade of innovation and creativity. From building trust and freedom to developing bendable organizations to tantalizing and arousing the whole person, the organizational setting provides a wallop in promoting creative thought.

Each one of these breakthrough creative elements has many

facets. Each is a center of energy. Each can produce much positive or negative force. Each can become flat, stale, limp, or lifeless. To produce maximum imaginative performance, you need to allow them to become an interactive whole. Then you can cause fluctuations in status quo thinking, unleashing ideas that give breadth, find a better twist, or discover a novel path.

The remainder of the book examines each breakthrough creative element in turn. We'll start with the potency of the human instrument.

2   The Human Instrument: Knowing Your Mental Equipment

> *Man's mind, stretched to a new idea, never goes back to its original dimension.*
> —OLIVER WENDELL HOLMES

Adventures and feats of science and engineering over the centuries have required a wide array of instruments.

Ancient astronomers used astrolabes—flat, circular maps of the stars with crossbars—to determine the location of planets and stars and to set the time and latitude. The development of elaborate sextants provided the next step in astronomical observations. Telescopes then took the scientist many leaps beyond those instruments that relied solely on the naked eye. Today's mass spectrometers accurately measure components of rocks taken from the moon! Astrophysicists, armed with supercomputers, can peer billions of years backward or forward in time!

Each field of inquiry develops instrumentation to carry out its work.

Another device, one that sometimes gets overlooked because it doesn't appear to have the precision of an ohmmeter or electron microscope, is the human instrument. The human apparatus can not only calculate and observe, but can also propose novel approaches, rehearse scenarios in the mind, or piece together a broad picture without all the details neatly in place.

You can develop yourself mentally and use more of your brain and mind power for breakthrough thinking by knowing about and using more of this exceptional instrument. If you work within groups, you need to allow the full mental functioning of all members of the team. Start with a sound understanding of yourself and what makes you both common and unique. Then apply that knowledge to an expanded creative interaction with your colleagues.

Look at some of the basics of your human equipment. When you are presented with a problem at work, your mind searches your memory for an approach you've used before. Or you scour past experiences for solutions that might work. You determine if you need more information or can go with what you currently know. You plug in an algorithm or set out to discover more data or answers. At some point you declare you have the solution and take some action to implement it.

The elegance of the solution will depend on how well you use your focusing and your mental processing equipment. Ordinary outcomes generally arise when you have not used all the possibilities of the human instrument. Breakthroughs arrive, not only when you have more mental channels open, but also when you access both rational and arational modes of thinking.

In this chapter, you'll be able to see how your level of consciousness, style of thinking, and modes of expression affect your creative output. First look at the basics of consciousness. Then explore how you process information, processes triggered by sensory input, flashes of memory, or emotional reactions. Finally, examine how you express your responses to incoming data. Once you are fully aware of your basic mental equipment, you'll be able to fine-tune it for optimal original thinking.

CONSCIOUSNESS: FOCUSING EQUIPMENT

While you read this section of the book, you are likely to be shifting your level of consciousness. That was true earlier, too, but now that it is being called to your attention, you will be more aware of the shifts.

At all times you are being bombarded by numerous stimuli. You are choosing where to place your attention—perhaps on a word, a cough, pressure on your legs or back, the sequence of a sentence, humming of fluorescent lights, a chair squeaking. Some of these stimuli are on the edge of your awareness. Various parts of your body automatically respond to the stimuli—orienting you to a new sound, for instance, then letting it fade into the background, or alerting your body chemically that food will be needed soon.

Consciousness, according to noted brain researcher and theorist Karl Pribrim, is *"what we pay attention to."* As you are reading along and your eyes catch a special phrase, for example, you may suddenly become fully involved in a fantasy event with your lover before realizing that your eyes are no longer skimming along the page. Your awareness has moved. Consciousness is where you put your attention.

"That's interesting," you say, "but what does that have to do with how creative I am?" Considerable. Our consciousness is often stuck on one channel. We stay on the level where we are most knowledgeable or comfortable or best trained. This creates blinders that can prevent us from "seeing," as Sampson Raphelson noted, "what's really there."

Galileo's improvements to the telescope helped us see more clearly in space. His calculations and experiments were a great boost to astronomy. Yet his level of attention was stuck on his belief that bodies in space travel in perfect circles. He therefore discounted comets as optical illusions.[1]

The world in which we live and move in varying states of consciousness has three arenas—inner, outer, and planetary—which serve as focal points for our attention and subtly, even unconsciously, affect us. Each arena includes stimuli, interactions, and known or unknown psychological, physical, and physiological realities. When we don't pay attention to these areas or toss them off as illusions, as Galileo did with comets, our mental vision has holes or blurriness. Creative thinking requires us to pay attention to a wider range of stimuli floating around in our mind-space. For maximum use of our imagination, we need to shift our consciousness readily.

What follows is a description of the elements that shape our attention in each arena.

Inner World

The *inner world contains sensations, thoughts, images, memories, emotions, and mechanisms for creating options*. This arena may be in or out of our consciousness.

When you sit in your office with your eyes glazed and your mind far away, a daydream can whisk you to the mountains to trek up a glacier or have you receiving acclaim for a new invention that sweeps the nation. You are focusing on the inner world of imagination. After a fitful night of fretting about a problem at work, you stand in the shower in the morning, enjoying the stimulation of the water on your body. Suddenly you think of a new way to handle the problem at work. The processes of intuition were out of your conscious mind. The attention was internal.

This inner world has *mind-sets*, or *ways of organizing the things that happen to people and the world*. These mind-sets are colored by myths, blind spots, values, world views, self-fulfilling prophecies, expectations, and images of who you think you are or want to be. They are shaped by present and past experience.

When you experience frustration over the inability of your work team to find a way through a puzzling situation, it is likely that you and your team are locked in and blinded by your collective and individual mind-sets. The mind's method of putting data into chunks for digestion, while helping you process much information, can distort, blind, or keep you from seeing things in fresh ways. For thinking creatively, these old mind-sets need to be broken so you can see things anew.

Outer World

The *outer world contains people, activities, work, events, and environmental influences*. It includes such external imagination stimulants as art, literature, music, cinema, video, group synergy, and the beauty of nature.

If you are engaged in a lively discussion with colleagues while trying to isolate the cause of an assembly line malfunction, your focus of attention, of course, is external. Your attention can switch rapidly. While you are debating with this task team, someone's stern tone of voice could remind you of a recent spat with your spouse. Your focus may linger internally for a while as you rehearse what was said and try to find a way to reconcile your thinking at home. Further comments by team members may cause you to shift your thinking back to the external exchange. Because of the internal distraction, you have missed a key point.

Planetary World

The *planetary world contains, metaphorically and dynamically, broader connections with Mother Earth, ancient patterns, all living creatures, and a collective unconscious.* Our planet, according to the theories and studies of a growing number of scientists, is more than rocks, water, gases, and growing things large and small. We may take planet life and matter so much for granted that we are oblivious to unobvious forces and forms.

Nobel laureate Prigogine, you remember, looked deeper into the planet. He saw structures invisible to the naked eye, forms exchanging energy, causing instability, and creating new patterns.[2] *Swiss psychologist Carl Jung noticed recurring themes in the dream expressions of people in urban Europe, the jungles of Africa, and India. To him, people were connected by a collective unconscious.*[3] *Rupert Sheldrake, a biologist, posed a controversial but beguiling theory. He suggested and tested the idea that there are invisible structures that communicate information across time and space—in plants, insects, people, animals.*[4]

A growing number of theories and experimental data point to possible interconnections of energy and information on this planet. We tend to discount the unseen and unaccounted for—just as former generations couldn't comprehend microscopic worlds that the electron microscope has delivered to us. The truly creative can tune their mental equipment to levels of consciousness oblivious to others.

In the early 1960s, MIT researcher Edward Lorenz studied the weather's unpredictability. To his amazement, as he was poring over his computer simulation, Lorenz discovered order amid the chaos. This seemingly disordered planet has remarkable patterns in its weather, geology, and chemistry. Computers, mathematical equations, and photographs have now verified these orderly forms. James Gleick's Chaos: Making a New Science *tells a fascinating tale of this new field of study.*[5]

The planetary arena can tantalize and push the boundaries of your thinking. While you hike through the woods, for example, your thoughts about walking on the crust of the earth could begin to broaden the horizons of your thinking. You might feel connected with that which surrounds you—majestic trees, a cascading stream, microcosms underfoot, animals in the forest, stars and planets hidden by a blue sky.

Critics might call that mystical. Some might say that's worship. Others might insist you've broadened and enriched your ability to observe, to be alert, to search for novel combinations.

Your consciousness probes for connections among the unseen. This is the kind of focus often required to "see what's really there" with your mental eyes without being able to see it with your sensory eyes! If people are to think with their full creative capacities, they must be able to spread their attention throughout the inner, outer, and planetary worlds. Then they are better equipped to explore and make discoveries in each arena.

THE BRAIN AND MIND: INFORMATION PROCESSING EQUIPMENT

As noted earlier, your mind and body are continually bombarded with stimuli. A pat on the shoulder. A car backfiring. Images on a television screen. An apple pie in the oven. The climax of a novel. Around the clock countless bits and pieces of information are processed in the brain.

You process this information in a variety of ways. You take in data with your senses. You sort and categorize it. You respond with thoughts, feelings, and expressions. Each person develops preferences or a unique style for handling these perceiving and reasoning processes.

These habits of the mind exert control. They govern where you put your attention when confronted with problems, the viewpoint you take toward finding solutions that are practical or novel, or the creative strategies you are willing to use. They control how you involve others in generating options and how you go about trying to sell and implement ideas.

Research into psychological types and into brain-mind functioning helps us understand modes that decrease or increase the power of imaginative information processing. We'll look first at psychological types, then at modes of the brain-mind. And finally we'll explore the modes and media that provide the greatest probabilities of enhancing creative thinking.

Psychological Types

There is a very useful model for understanding how you handle information that confronts you. According to the theory, you have tendencies to lean toward certain types of behavior and thought.

A scientist who likes to dig quietly into research on her own may enjoy spending days in the lab squinting through a microscope. The entrepreneur continually looking for new products to manufacture gets charged up trying to sell his latest idea to top management. An engineer teaching college-level courses revels in the chance to inspire and challenge each new crop of students.

You, along with these three representative people, are guided in daily transactions, in choosing a career, in conflict situations, and in problem solving and decisionmaking. Your guide is an internal steering mechanism. You have programmed this device by the modes of thinking and behaving you prefer. Since you were a wee one, you have been shaping and are being shaped by a combination of psychological preferences.

The great Swiss pioneer of the mind, Carl Gustav Jung, proposed the idea of unique, personalized, psychological types. After conversations with Freud—his mentor, colleague, and friend—turned sour, Jung solidified his thinking about the psychological positions or types people have. He reasoned that these types may cause allure, puzzlement, resonance, comfort, or strangeness when interacting with someone of a similar or dissimilar type.

After studying Jung's seminal work, Isabel Briggs Myers used the theory of psychological types to develop an instrument that assesses one's type. The test, the Myers-Briggs Type Indicator, has been researched considerably during the past thirty-plus years. It is a very beneficial tool for understanding what makes you unique, how you are gifted, and where some of your flat spots might be.

You won't be taking the test here, just determining the degree to which you believe you fall more onto one side or the other of four dimensions. It won't be as accurate as taking the test, but it'll give you some indications of what type might fit you. These descriptions should give you some notion of your strengths and how others might see your flat spots. No one is a totally lopsided individual. You will be able to see parts of yourself in both camps or poles. Based on the descriptions here, however, force yourself to choose your preferences.

If you don't have anyone in your company who can give you the full Myers-Briggs Type Indicator, buy the popular paperback *Please Understand Me* by David Keirsey and Marilyn Bates. The authors have a short test in their book that can give you the same psychological types. The book is a winner for learning more about how the types apply to your work, your creative endeavors, your home life.[6]

Characteristics of the psychological types, with strengths and cautions for creative thinking applications, are as follows.

Outer-Inner Focus

Extroverts have an interest in the world outside them, where the focus is on people and action. Their energy expands through interaction with others. They think as they talk. Generally they have broad interests.

- **Strengths:** Able to problem-solve in groups. Willing to express thoughts and ideas, to act and interact.

- **Cautions:** May overpower and inhibit introverts, expecting them to express themselves before they have had a chance to think things through internally. Might not take the time to reflect on the ramifications of an idea before leaping into action.

Introverts place their interest and energy in internal thought processes. They move their energy away from the environment and focus it on ideas and concepts. The introverted think, then speak. They delve deeply into things that interest them.

- **Strengths:** Willing to wrestle with topics in depth. Want to understand the world. Reflect, then talk.
- **Cautions:** Often need private thinking or writing time before expressing themselves. May not readily take counsel from someone else. Other people can drain their energy. If they don't share the thought processes they are experiencing internally, they may mislead a person who is only privy to their words and nonverbals.

Perceiving Data

Sensors take in information based on its practicality, impression on the senses, orientation to the present, and realism. They are interested in facts. To be real, data must be experienced by the senses—seen, touched, tasted, heard, smelled. Therefore, sensors often live life as they experience it, moment to moment.

- **Strengths:** Generate practical ideas and action steps when originality isn't needed. Come up with facts when analyzing problems. This here-and-now orientation is especially useful for artists and creative entertainers who must experience and express themselves in the present.
- **Cautions:** This practical bent may prevent sensors from seeing the possibilities in imaginative options. May be impatient with the number of ideas intuitives want to generate. May not see the big picture.

Intuitives perceive data based on its possibilities. They tap a process out of their awareness, responding to hunches, images, a "gut feeling." Their focus is on the future. Using their imagination invigorates them. Complexity whets their appetite. Change keeps them fresh, letting them try out an ongoing stream of ideas.

- **Strengths:** Like to search for possibilities, stretch their imagination, solve complex problems, see the big picture, and probe the future. Novelty is seen as important.
- **Cautions:** May push for ideas beyond what is needed for the situation. Might neglect facts, practical implications, and the present.

Making Decisions

Thinkers look for logical connections, thus making decisions based on careful analysis and impersonal arguments. They are often guided by principles and their perception of truth.

- **Strengths:** Analyze problems and evaluate options.
- **Cautions:** May retain an impersonal viewpoint when sensitivity to human values and emotions are required. May rule out intuitive thoughts, even their own, when the ideas don't seem logical. Might be critical of the ideas of others.

Feelers use a rational process to make judgments, but do so *by weighing values and individual or group considerations.* They base decisions on personal values. These people often observe needs, emotions, and communication cues that can be factored into the decisionmaking.

- **Strengths:** Add an orientation to empathy, people's values, emotional color, and group maintenance in team sessions. Provide enthusiasm and enjoyment.
- **Cautions:** May be uncritical, illogical, or unorganized at times. An overfocus on the personal may mislead when the real issue is technical.

Outer World Functioning

Judgers organize and structure the world around them. They plan, are decisive, like to be right and exact. When confronted with a situation requiring a decision, judgers don't care to dwell on gathering information. Instead, they want closure—on activities, ideas, and thoughts.

- **Strengths:** Provide discipline and structure. Willing to make decisions and act. Share their judgments.
- **Cautions:** Might be rigid and too controlling. May not like to make changes. Could be prejudiced with an underdeveloped use of perception. May resist revisions when events clearly indicate a need to make a change.

Perceivers keep things flexible, spontaneous, and ready for change. They like to keep the door open to new data and dislike being pushed to make conclusions before they are ready.

- **Strengths:** Willing to explore, flow with a process, or adapt. Open to new information.
- **Cautions:** May delay decisionmaking. For fear of being tied to judgments they have made, perceivers may not share their decisions. Their desire for openness to new data may cause others to feel the situation is out of control. Could procrastinate if they have an underdeveloped sense of judgment.

Any combination of the four dimensions is fine. Each of the sixteen possible combinations can enhance the synergy, variety, and development of ideas. Each "type" blend has important contributions to make or strengths to give to creative endeavors. Each has its rough spots and blind spots, its strengths that are overused.

Knowing your type is not sealing your personality in a straitjacket. Becoming aware of your psychological tendencies makes you more valuable for creative thinking. This knowledge gives you one more perspective on how well you are capitalizing on your strengths, minimizing the flat spots, and developing parts of you and others that are underused.

Knowing your psychological type is part of the picture. Knowing your reasoning style adds more to your portrait of the mental-emotional-perceptual equipment you have. We turn to thinking patterns next.

Brain-Mind Modes

What do you gain from studying how the brain and mind interact? You get clues about what has to happen to unleash your imagination.

Exciting hypotheses were made in the 1960s and 1970s about the potential of the brain to think creatively. The right hemisphere was trumpeted as the place in the brain where imaginative reasoning takes place. Some educational and psychological pundits made zealous claims that caused people to think too simplistically about the workings of the brain and the seat of creativity. However, much has been learned about human imaginative abilities, especially the modes of processing information that lead to or detract from innovative thought. Tracing the history of this brain-mind study may help show the mental processes that aid creative thought.[7] The following summarizes that history.

An enormous amount of research was touched off by effects of the surgical separation of the two hemispheres of the brain. Starting with monkeys in the 1950s, Roger Sperry severed the corpus callosum to study the processes of each brain half more carefully. He cut the connecting tissue and neural pathway between the hemispheres. He conducted the operation on severe epileptics in an attempt to reduce their debilitating seizures. Each hemisphere was able to operate independently. Not only could they think separately, but the brain halves used different methods to process information.

During the 1960s, neurosurgeon Joseph Bogen named the hemispheric tasks and linked creativity to the right brain. The left brain, he said, was adept at speaking, calculating, and reading, while the right hemisphere specialized in faces, spaces, and mazes.

In the 1970s, another noted brain researcher and neurophysiologist, Karl Pribrim, was puzzled by the ability of brains damaged by lesions to recover. He conjectured that the damaged brain recovered memory because of common memory holograms distributed throughout the brain. At first he used the hologram as a metaphor, then began to discover physical evidence to back up his theory. David Bohm, a noted London physicist, was on a similar track, and together they were able to provide data to back up the "holograms-in-the-brain" theory.

In the recent past, there have been challenges to the old findings and new suppositions about the functioning of the brain-mind that bear on creative thinking:

- The hemispheres work in a complementary fashion. The left hemisphere is good at detailed processing, use of language, and fine analysis. In the right brain, there is more proficiency in larger, broader processing, with better interpretation of vague, novel information. Spatial relationships and global analysis are also right hemisphere specialties. The right-brain half appears to have closer emotional and intuitive ties.
- Psychiatrist William Gray has theorized that thoughts aren't just intellectual, but are embedded in emotional tones or codes. One gets an *aha* when feeling codes or knowledge that have been incubating in the brain interact with feeling tones that arise in the midst of a new experience.[8,9]
- Combining senses and modes of thinking enhances learning. The *Brain/Mind Bulletin* has cited a number of studies showing the connection. The combining of relaxation, imagination, and concentration techniques, for example, has doubled performance on writing and speaking tests.[10] In another study, strong visualization and relaxation led to increased performance.[11]

A key for understanding creative thinking in all this research is that there are styles of thinking that can enhance, dampen, or complement imaginative reasoning. These modes of processing information are very important. Yet the physiological seat of a mental task as complex as creativity hasn't been pinpointed. Because creative thinking requires a combination of attributes and mental processes, we will likely never be able to point to a spot in the brain that is the genesis of creativity.

Again, the reason I am exploring the brain and mind is to increase awareness of how we process information. Our minds are often taken for granted—silently churning along, seemingly at our beck and call for automatically grinding out answers, facts, memories. Yet we have far more control than we often realize about how we take in data, turn that information around to see it from different angles, and make conclusions about it.

Rational and Arational Modes

For the sake of comparison, the brain reasoning activities can be split into two distinct modes: the rational and the arational (see Table 1). In actual working, however, there is much interaction, much blending between these modes. Everyone uses both daily. As with psychological types, however, people often prefer one style. An individual, therefore, tends to use one mode more than the other.

If your *rational mode* is predominant, you try to reduce or squeeze data down to their essential elements. You sift through a variety of information in a step-by-step fashion, looking for logical connections. Time is taken to weigh bits and pieces of information. Through analysis of the data you arrive at a conclusion. Your primary vehicle of analysis and expression is verbal—using known symbols of language and mathematical and scientific equations that are familiar to you and to others with whom you want to communicate.

Table 1. Rational and Arational Modes Contrasted

	Rational	Arational
How you think	Reduce Step by step Analyze Use verbal	Expand Simultaneous Intuit Use nonverbal
What you do	Search for correct way	Search for variety of ways
What this requires	Precision Control	Approximation Letting go
How you respond	For arational thinkers, this seems confining, slow, narrow, detailed, unnecessary, too linear	For rational thinkers, this seems chaotic, diverting, scary, obscure, untested, too random or risky

In the rational mode, you are seeking the correct approach or answer. Precision in making comparisons and calculations is vital. Therefore, you seek to control the data so you don't add extraneous factors to your analysis and so you can track the many pieces in a coherent manner.

Following a rational path may be critical in completing many engineering and scientific tasks. To the arational thinker, however, this rational approach often seems confining, slow, narrow, detailed, unnecessary, or too linear. Emotionally and mentally the arational reasoner would feel bound in—not as imaginative or productive as the person would like to be, not as able to arrive at breakthrough thinking.

If your *arational mode* is on center stage, different mental functions stand out. You allow the data to grow, to expand into whatever patterns emerge. You take in information from a variety of sources or directions simultaneously, using a soft rather than hard, rational focus. Intuition is the key mechanism that helps you see a bigger picture from pieces of information, from clues and cues that give you an image, impression, or sense of the answer or direction to be taken. Conclusions may come in the form of images, sounds, or feelings in your body.

In the arational style, you want variety rather than correctness, approximation rather than precision. Getting close allows you to have the opportunity to shape ideas to requirements, rather than insisting that a thought or idea be perfect on the first try. The arational style requires you to let go of your need to massage and control the data to fit a fixed template, to let go of a favorite idea or of a stereotype, to release a mind-set.

Rational thinkers often feel the arational approach is chaotic, diverting, scary, obscure, untested, or too random or risky.

Both the rational and the arational styles are crucial for clear thinking, with the arational style being critical for creative thinking. Our current educational system doesn't accentuate the dual modes. Arational thinking is often stunted through lack of attention and expression. To provide more of a balance to the rational style that is so much a part of the way we have been taught to think, this book stresses the arational mode.

PROCESSING AND EXPRESSING CREATIVE THOUGHT

Within the rational and the arational are special ways we process and express creative thinking. Rational thought is represented by language and the use of well-known symbols to convey reasoning verbally. Arational thinking has a wider array of processes and media. Intuition and the preconscious are arational processes. Vision, audition, and the kinesthetic sense are arational media.

Figure 1 shows the relationships between arational processes for tapping creative thought and the media for expressing or transmitting ideas. Linked together, language, as the rational medium, and sight, sound, and touch or movement, as arational media, provide powerhouse vehicles for expressing innovative thought.

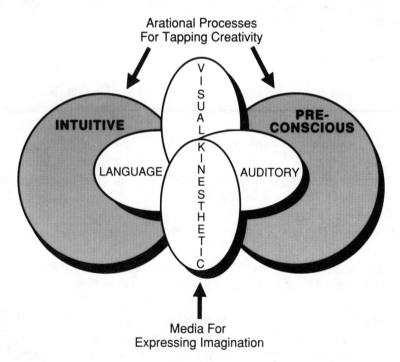

Figure 1. Expressive media and arational processes.

When you use rational and arational styles in a combined, interactive way, you can have balanced, productive, and imaginative reasoning. If, however, you overwork the rational mode, you concentrate on language and may neglect sensory expressiveness or thinking, as well as intuitive and preconscious processes. Or, if you get locked into one arational medium, such as visual thinking, you can inhibit the full use of auditory and kinesthetic media.

Stethoscopes, statistical software, and microscopes are only a few of the many external tools available to help your rational thinking. You also have potent built-in equipment for generating arational or creative thoughts. Much rests on how well you maintain and develop these internal creation processes and media.

Primary Arational Processes

Mechanisms responsible for arational processes are intuition and the preconscious.

Intuition

Intuition, the central process for producing imagination, is a mechanism of direct knowing. Without labored analysis, you just "know" something is correct or the thing to do.[12] Colleagues, specialists, or the troops may say the right way is the other way.

Market research, for instance, said the public wouldn't accept a revolutionary new camera. Inventor Ed Land intuitively "felt" the public would buy it—and pushed through the development and marketing of the vastly popular Polaroid.

There are countless other examples of intuition's hand in the work of researchers and scientists over the centuries.

Michael Faraday, one of the 1800s' most remarkable thinkers and contributors to knowledge, had a very scanty mathematical background. Yet his intuitive experiments and puzzling over natural phenomena uncovered important properties of electricity and produced the dynamo, in which mechanical energy could be changed into electrical energy.

Alfred Wallace pursued some of the same questions as his friend Charles Darwin. Wallace was struggling to find the key to what he, like his friend, believed to be the evolution of species. He was struck with the idea that species developed in a wide variety of environments, some favorable, others hostile. Yet he pondered over one of the pieces of the puzzle, the fact that organic life coming from the same stock could become so diverse. While wrapped in blankets to ward off chills from a bad fever, Wallace's mind wandered to Malthus's An Essay on the Principle of Population, *which spoke of increases in population being a cause for decline, not improvement of species. An aha struck him. His bolt of intuition was that only the fittest survive! He spent several feverish evenings writing down his theory to send to Darwin, thinking the idea would be new to his friend. When he received the letter, Darwin, who had come to the same conclusion some years before but hadn't published his thoughts, was struck by the coincidence in their thinking. He, too, had been influenced by Malthus in coming up with the key of the survival of the fittest. Darwin, by dint of dogged pursuit of the theory, became the one most remembered for the evolutionary principle.[13]*

Many people wonder where they got their hunch, the thought to try something a new way or to take a new approach. A number of researchers have hypothesized that the right brain may be the source of intuition. The intuitive mode seems to have a kinship with sight, hearing, touch, smell, taste, emotion, experience, memories, various levels of consciousness, and processing of information that often captures the whole picture. Or it seizes the essence of something and often delivers it to the conscious mind in images, fragments, or sensory awareness.

Many people cover up their intuitive thoughts by denial or by logical reasoning. Other individuals are very receptive to intuitive experiences and processes. A growing number of people in the business and scientific world are exploring the potential of intuition for providing innovative ideas and for making decisions.

Intuition as used here is the process itself, while *intuition* in the Myers-Briggs format is a preference for using the process.

Preconscious

The *preconscious is a level of consciousness just out of reach of your waking, hard-focusing awareness. It is an arena in which more primitive processes wrestle with and resolve problems.* The preconscious comes into play when you aren't fully awake or when you are actually asleep. It is a time of altered brain waves and activity. As you begin to relax or shut out the clamor of stimuli in your waking hours, preconscious mechanisms appear to be able to creatively tackle your problems.

Especially valuable times are during presleep and during dream sleep, or during rapid-eye-movement sleep. Dreams occur approximately every 90 minutes. Shorter dream periods occur early in the night and longer periods of up to 20 minutes take place toward morning. During night dreaming and while falling asleep or waking up, creative people have made discoveries or found clues to puzzling problems.

When August Kekule fell asleep in front of a fire in the winter of 1865, a dream led to an important discovery. Read his description, as related in *Creative Dreaming,* by Patricia Garfield:

> *Again the atoms were juggling before my eyes ... my mind's eye, sharpened by repeated sights of a similar kind, could now distinguish larger structures of different forms and in long chains, many of them close together; everything was moving in a snake-like and twisting manner. Suddenly, what was this? One of the snakes got hold of its own tail and the whole structure was mockingly twisting in front of my eyes. As if struck by lightning, I awoke....* [14]

From the clue in the dream, he built a hypothesis: the benzene molecule had a structure like a ring! It opened the door to understanding many other complex chemicals. Today we easily accept the notion of a closed molecular structure. It wasn't always so, though, and took the aid of Kekule's preconscious to make the critical discovery.

While a medical resident in 1975, Frank Barr had a series of ten dreams he described as bizarre. One of the strangest features of the dreams was a blackness that he experienced as being in his brain. Intrigued, he began to investigate this special blackness. His investigations led him to propose a comprehensive theory about the nature of the energy exchanges in melanin, the important coating around nerve cells. To this day his views are commonly accepted.[15]

In her *Creative Dreaming*, Patricia Garfield described the remarkable 1893 dream of Hermann Hilprecht, a noted anthropologist at the University of Pennsylvania.

Hilprecht was trying to decipher the fragments of two drawings that had come from temple ruins. He thought they might be rings. The anthropologist temporarily assigned one fragment to a specific time period, but he couldn't figure out the other one.

He had a dream in which he was told the secret of the two fragments. A temple priest appeared in his dream, ushering him into a temple chamber. Then the ancient man described to Hilprecht how the fragments really belonged together. The priests had received a command that they were to make two earrings for the statue of the god of Ninib. The message to them, however, was that the earrings were to be made of agate. To the priest's distress, there was no agate readily available. They, therefore, took a previously crafted gift, an agate cylinder that carried an inscription, and cut it into three rings. Two were used as earrings, while the third ring disappeared.

Hilprecht awoke and told his wife. In the morning, he studied the drawings, verifying the inscription that could be understood when the two rings were paired. Later, when he was able to travel to a museum in Constantinople, he located the two rings being kept in separate displays. They indeed had come from the same agate cylinder.[16]

Exactly how the preconscious works its wonders in dreams is not known. It could be a time when the mind tries to pool and make sense of data in the memory bank. It could be the same processes that work intuition. It could be random firing of the mind to clean out the clutter for the next day's work. Whatever the mechanism, the preconscious creates a treasure chest of new combinations and possibilities, if we are willing to enter its chambers and listen to its wisdom.

Media for Expression

Media that express creative thought also carry imaginative potency.

You may not have spent much time in the past trying to figure out the metaphors and patterns you use to express yourself. Yet you do have unique ways of gathering information. And you represent experience in distinctive visual, auditory, and kinesthetic ways. Your language carries the imprint of your expressive style as you use symbols like words, equations, and formulas to convey ideas.

John Grinder and Richard Bandler, coauthors of *The Structure of Magic II*, are two recent students of communication patterns that focus on the impact of the media people use to convey thought (neurolinguistic programming). They highlight the visual, auditory, and kinesthetic senses and exclude taste and smell as not playing a major role in representing experience. Smell, however, has been identified in other studies as the sense that often brings the most emotional sensitivity or intensity.[17] The author of *The Metaphoric Mind*, Bob Samples, has also eloquently shown the power of the media in the expression and stimulation of imagination. A summary of these media follows:

Visual

The *visual medium is a combination of the visual sense, the system it represents, and the image-making capabilities of the mind*, with exterior and interior dimensions. The world outside the individual can be brought within through the eyes, but marvelous images can also be painted on the mind's eye—without opening an eye.

One of the marks of creative people is a capacity to create vivid images. You can improve your imagination by developing your ability to see more acutely—to "see" the smells, emotional tones, nuances, colors, subtle movements—either in the external

THE HUMAN INSTRUMENT: KNOWING YOUR MENTAL EQUIPMENT   47

or in the internal worlds. You can increase your skills in expressing your thoughts in images on paper, in the mind, and to an audience.

For many years, the ability to think visually was felt to be insignificant by psychologists, who thought only external behavior was valuable. But research and new applications have brought a resurgence of the powers of image making. The ability to visualize is seen as an ability to:

- Create specific new combinations and mental manipulations
- Tap intuitive and preconscious thought
- Extract essence and the big picture
- Draw on memories and emotions
- Rehearse and produce successful performance
- Improve memory
- Alter bodily responses
- Reduce fear

The art of invention involves great skills in visualization.

Michael Faraday, for instance, had an exceptional ability to see in his mind the curved lines created around electrical currents and magnets. The lines in space were real to him, not metaphorical. His internal vision allowed him to make significant contributions to the field of electricity.[18]

Where can you find image making that can trigger, produce, and express imagination? An excellent start is to form metaphors and analogies. They are word pictures that can help you make fascinating, useful connections between objects or events in the world around you and the problem you want to solve in a new way.

Exercise *the visual: Use your internal camera. Create scenes on your mental screen. Let mind movies reconstruct experiences from your memory bank. Or construct an invention. Turn it around. View it from different directions. Cut away what you don't like.*

48   CREATIVE THINKING AND PROBLEM SOLVING

Expand or alter what you do like. Mental simulation is very cost-effective.

Track your mental camera when you are very relaxed or asleep. Intuitive, preconscious processes are more active then. Observe the imagery just as you are ready to fall asleep or as you wake up. Note your dreams. You may even become expert at altering the plot or images.

Create external visuals. Keep a pad by your bedside. Carry a journal in your attache case. Have it handy at your desk. Then draw. Sculpt. Develop a prototype.

Sketching your mental images is an excellent way to make your visualizing real. Draw various versions of an idea—a device, process, method, or theory. Roughing out the mind picture in your journal or on a note pad captures it, holds it for your use at a later time. Freehand drawing is fine. So are stick figures and crude sketches. You aren't vying for the artist-of-the-year award. You are trying to grasp a concept.

Edison made many crude drawings, sketching to convey ideas. But he was also wise to the value of having a recording for patent development. A few lines on paper is what Edison handed to his machinist. It was his simple vision of a phonograph (see Figure 2).

Leonardo da Vinci, was a prolific sketcher. Granted, he was a memorable artist. But he made countless drawings that weren't intended to be works of art. His sketches were his way of getting the devices and concepts dancing in his head into the open where he could play with them. His sketch of a two-level bridge is shown in Figure 3.

Auditory

The *medium of hearing* is one of the primary senses, the *ability to perceive and make sense of sounds*. Most people take it for granted, along with the beating of one's heart and everyday noises. When sounds are soothing, you can relax or be stimulated. When sounds are jarring, you can get nervous or, if loud enough, feel

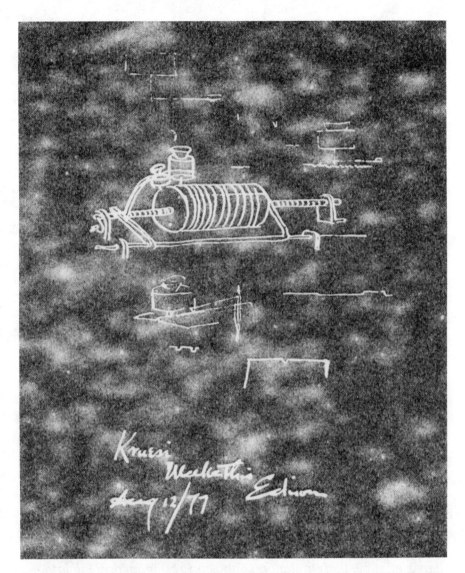

Figure 2. Edison's concept of the phonograph. Reprinted by permission of U.S. Department of the Interior, National Park Service, Edison National Historic Site, West Orange, New Jersey.

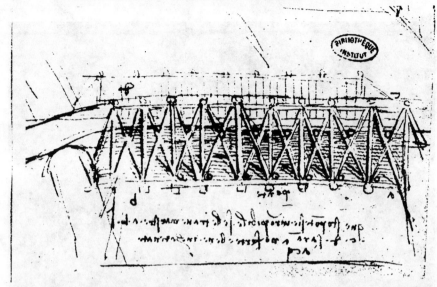

Figure 3. Leonardo da Vinci sketch of a two-level bridge, with an upper level for pedestrians and a lower level for vehicles. Reprinted by permission of International Business Machine Corporation, New York, New York.

pain. Sounds and words coming to you through the radio can stir the imagination.

In a famous radio spot, comedian Stan Freberg extolled the virtues of radio as an excellent vehicle for stimulating the imagination. He said there were things one could do on radio that couldn't be done on television. To make his point, Stan commanded a crew to roll a 700-foot mountain of whipped cream onto Lake Michigan, which had been drained and filled with hot chocolate. The Royal Canadian Air Force dropped a 10-ton maraschino cherry onto the whipped cream on cue as 10,000 extras cheered wildly! Of course, it was all done with imitation sounds and by triggering the imagination.

Some people are especially tuned to sounds. Musical scores can be played in their entirety in the heads of some individuals, a trait that is very handy for composers—or those who want to score big on a "Name That Tune" television program. Certain mechanics can "hear" the problems within a motor. People who are auditory thinkers will often sprinkle into their expressions such words as: "I hear you." "There's a rhythm in my work." "What does this proposal sound like to you?"

Exercise *the auditory: Use the auditory medium in different ways. Explore a topic by asking, "If you could express this problem as a sound, what would the sound be?"*

Or you can use sound to create a mood that brings relaxation and stimulates your intuitive processes. The stimulative power of music varies with individual tastes. However, music that is slow baroque, approximately sixty beats a minute, and 4/4 or 3/4 time seems best for producing relaxed concentration. A range of instruments can be good for balance, but lyrics can be distracting. Environmental sounds—waves rolling into shore, birds calling, wind whistling through the trees—also offer reflective sounds. Use relaxing music for individual or group sessions in which you want to stimulate creative thinking.

Kinesthetic

The *kinesthetic medium is composed of sensitivity to touch and body movement.* It is one of the first ways we learned about our environment as children. The kinesthetic, as one of the key media, represents the often subtle but highly important connections between the mind and the body. Einstein, for instance, claimed that he had a feeling of muscular involvement as his ideas developed. He was a kinesthetic as well as a visual thinker.

Exercise *the kinesthetic: Here's a kinesthetic thinking technique you can try. Introduce to your mind the problem you have been trying to solve. Take a long walk. After you have developed your walking rhythm, start talking out loud about the different factors that must be considered. You might try this first with no one else around. Allow yourself to produce and describe a host of ideas that could be tried, including wild ones that you think no one else would think of. Match the vigor and pace of your stride with your words. Put more emotional color in your movements as you try to express an idea or search for one. You can return from a walk like this with new perspectives and fresh ideas.*

Language

The *language system uses words and other common symbols to communicate*. It is the medium of the language center and is strongly tied to logic and analysis. Your words can be used to gain precision, as in the well-chiseled words of a theory. But your words can also expand and lift when formed into a poem or the lyrics of a song.

Exercise *the language medium: Expound upon a subject. First use crisp, precise words. Next, amplify the topic, but wax poetic, letting your imagination and spontaneity free. Use word pictures and images. Try the second method to loosen the grip of the overly rational. Or see how many unusual interpretations you can make to common symbols, like a stop sign or familiar equation. One of the techniques Edward de Bono suggests is to pick a neutral word, then see how many associations and ideas you can create around that word. Go to another word, letting imaginative thoughts build again.[19] This technique can help you move beyond the literal and concrete, allowing your intuition to grow and your sense of new possibilities to be enhanced.*

SUMMARY

The human instrument can be a forceful cluster of energy in the work of scientists and engineers. The basics may seem too basic. We wear our styles of processing information or our psychological preferences comfortably and often unconsciously. Our styles have been with us even longer than that old pair of shoes that we hang on to for years because they have been so comfortable or because we are too sentimental to give them up.

The basic human equipment is present at all times, but it can get out of sync or lose its calibration. We need to fine-tune ourselves to more readily move to an appropriate level of consciousness for the task, to blend the rational with the arational, and gain quicker access to our intuition and preconscious.

We will now turn our attention to those qualities of the human instrument that go beyond the common and the basic to promote breakthroughs of creative thought. Knowing and exercising our mental apparatus is a fine start. But if we also master the breakthrough qualities in the next chapter, we will have the mental tools to embark on the journey toward *eureka*'s that will fill the rest of the book.

3 The Human Instrument: Expressing Breakthrough Qualities

To make headway, improve your head.
—B. C. FORBES

It is possible that scientists, poets, painters and writers are all members of the same family of people whose gift it is by nature to take those things which we call commonplace and to represent them to us in such ways that our self-imposed limitations are expanded.[1]
—GARY ZUKAV

For creative output, the human instrument needs to be in peak operating order. Yet your equipment for making discoveries and unleashing imagination may be idling, lax, dormant, blocked, or out of kilter. Your level of consciousness may be too narrow, thinking preferences too confining, or reasoning modes and media too lopsided.

For the human instrument to operate with innovative effectiveness, certain qualities are needed. The first quality, *soft focus*, allows people to handle more data, see more options, and not be hemmed in by constraints. *Childlikeness*, the second attribute, permits individuals to be fascinated with the world, explore ideas with curiosity and playfulness, and be emotionally stimulated. Both qualities will now be examined.

BREAKTHROUGH QUALITY 1: SOFT FOCUS

An important element of your basic equipment, as discussed in Chapter 2, is your level of consciousness. For the purposes here, there are essentially two types of consciousness to be considered. While useful in some endeavors, the first method of focusing, hard focus, can stifle creativity. The other, soft focus, is essential for creative thinking.

Hard, Narrow Attention

The first level of consciousness is a type of *attention where you block out much of the stimuli available to you and have a hard, narrow focus upon some activity, thought, or goal.* You attend exclusively and with energy. The ambitious, hard-driving, Type A personality is a metaphor for this kind of consciousness. Actually hard focus is epidemic in Western cultures. Schools and businesses thrive on pushing this type of consciousness.

Lester Fehmi, a biofeedback psychologist at Princeton's Biofeedback Research Center, sees positive aspects in this narrow attention when a problem must be solved very quickly or when one is absorbed in eating chocolate or listening to a symphony. But, when it becomes rigid and chronic, hard focus can produce many of the symptoms of stress—headaches, ulcers, colitis, knots in the stomach, diarrhea, and fatigue.[2]

Seen in the perspective of imaginative thinking, hard focus can limit novel options by blocking out creative responses and inducing invention-killing stress.

Hard focus can produce mental blinders.

Place blocks tightly in a line. Leave space somewhere in the line for one block. Move one of the two blocks beside that void into the space. What moved, the hole or the block? Try moving another block, then another. The hole moved in the opposite direction from where you moved the block. A mental set develops from focusing on the block that was moved, not the space that was created.

For many years Edison clung to the use of direct current even though it was ineffective when carried more than two miles. He spurned the more efficient alternating current transformer invented by Nikola Tesla and developed by George Westinghouse and William Stanley.[3]

Another highly original thinker and father of the heat engine, Sadi Carnot, insisted heat was just a fluid. He wouldn't accept Joule's important experiments that showed that heat could be produced mechanically.[4]

This narrow type of consciousness afflicts the normally creative, as well as those with sparkling mental brilliance. Hard focus sets up barriers to open, flexible thinking. Leafing through the pages of scientific and engineering history, you can uncover numerous other examples. You might cite examples of how such narrow attention has hampered your work or that of your colleagues.

Soft, Open Attention

The second type of consciousness is a soft, open focus. *Attention is flexed, and the person is relaxed.* If you are in soft focus, you are open to the stimuli of the world. You can *often focus on the task and the surrounding environment at the same time,* integrating both.

Everyone has, in some measure, experienced soft focus. During a lengthy tour of a competitor's plant, for instance, you may notice a period when you are aware of your physical tiredness and also alive to the sounds and sights on every side of you. While communicating with a friend, you might experience soft focus by being fully alert to the flow and nuances of the conversation.

Soft, open focus is a style of consciousness that often accompanies optimal performance, whether in the arts, athletics, or business.

A juggler uses soft focus to keep an eye not only on the objects being juggled but also on the space between the objects. A fullback swerving and dodging his way up the field works best by paying attention to shapes, spaces, and colors, rather than the defensive linemen directly in front of him.

A top runner racing along a track will often be aware of her breathing, relaxed movement of her body, pace, the moving forms around her, feel of the track, and feedback of the crowd. The soft focus of the athlete can be beauty in motion, using mental grace and strength to be centered and flexible, while responding to the surroundings like an aikido master. A manager can orchestrate a project, combining diverse people and tasks harmoniously to arrive at a successful conclusion.

Soft focus is critical to optimal creative thinking, for it expands your options, is receptive to wider ranges of stimuli and the senses, and reduces stress that blocks effective, imaginative thinking.

Think Outrageously

With a more open focus, you can think outrageously! Sound strange? *Webster's Ninth New Collegiate Dictionary* has a definition that fits imaginative reasoning. It states that being outrageous is "exceeding the limits of what is usual." It is "not conventional or matter-of-fact." This is the kind of reasoning that doesn't get trapped by boundaries.

When you use your intuition to operate in an open focus way, you can see the big picture. A boundary just becomes part of the broader mental landscape. You blur the barriers; look right past constraints; see ways to go under, over, or around boundaries.

Did you gasp when you saw the suggestion to think outrageously? Many of you work in organizations with procedures piled to the ceiling. You are expected to be vigilant in following the rules. Guidelines about how to behave in many systems can be crucial for safe operations, consistency, and ease of doing things.

The repetition, the caution, the following of patterns can become habitual, however. Then, when the situation calls for us to solve problems creatively, we can be stuck in a do-it-by-the-book mentality. Just as there is a time to be rational and another to be arational, there is a time to follow the rules and a time to move beyond the constraints with the use of imagination.

If you are a strong intuitive type in the Myers-Briggs sense, you may just naturally push boundaries. You might treat rules of the corporation or work procedures or even the way problems are defined as something that may be challenged. You may want to change or modify or reframe any boundary condition that stands in the way of peak performance or of solving a thorny problem. It often takes bold, outrageous thinking to identify a new approach and champion it in a company. For some of you, outrageous thoughts are part of your nature.

Others of you may need to give yourself permission to treat a boundary condition as flexible and not set in concrete. Certainly there may be givens that are part of a problem solving situation, ground rules for the common or business good, and constraints that are real. Far too often, however, boundaries are more mental than true barriers.

There are blocks that can constrict your soft consciousness and limit your ability to relax and flex your attention. If, for example, you perceive a constraint to be fixed and immovable when there are ways around it, you are caught in a trick of the mind. Mind-sets can be deceiving. You can believe a wall has been erected when it is really an illusion. You can believe that the constraints you perceive are holding you back wipe out the chance for good solutions.

When young Philo Farnsworth proposed an idea regarding television broadcasting, many in the engineering community thought that such broadcasting was impractical. If Philo had agreed with the constraint of impracticality, the development of television might have been much slower.[5]

In the mid-1970s, there was a cartoon film about Joshua, a glob-like character who found himself living in a box. At first he was relatively contented, not realizing he was in a box. Once he realized he lived and moved inside the constraints of a box, he began to feel hemmed in. He fought and fought to get out of the box, but his angry blows were to no avail. Exhausted, he leaned against the box and found that there was a hole he hadn't seen. He squeezed through to the outside, then danced in celebration of being free. In time, Joshua began to get very lonely outside the confines of the box. Therefore, he built another box around himself, although the new container was just a little bigger than the old one. The box had been his own creation in the first place. When he built the new box, he pushed the boundaries of the box, so he had a bit more room and sense of freedom than before.

Many people build a box around themselves and set up boundaries that appear to be real. The hoax is usually self-imposed, however. Numerous innovations have been made to the incredulity of colleagues who believed a positive solution was impossible. A hard focus locks us in. A soft focus gives us a key for exploring creative possibilities.

Soft Focus Block Breaker

To break the hold of your hard, narrow concentration, focus on how you focus.

Practice flexing or softening your attention. Be alert to the space between objects and parts around and in you. Imagine the volume of the objects and parts—to see them in totality and unity—like a juggler absorbed in the space and volume of objects moving in space. Multiple balls in the air are seen as a unit with their surroundings.

Try this: *Relax in a comfortable position. To spread and soften your attention toward the world in and around you, begin to imagine, with eyes closed or open, the space or volume inside and outside you. Imagine the space or volume, for instance, between your eyes; between the top of your head and your jaw; within your chest cavity; surrounding the bones on your hands; between you and specific objects close or far from you in a room. Add more areas of your body and your surroundings. Practice 15 minutes per day. This flexing of your focus can be tackled in a disciplined way by using Fritz and Fehmi's The Open Focus Handbook and cassette tape program.*[6]

Practicing soft focus approaches can loosen the rigid grip of hard focus. You can then move more freely between an open focus and an appropriate narrowing of your attention.

BREAKTHROUGH QUALITY 2: CHILDLIKENESS

An adage making the rounds not long ago was: Real men don't eat quiche. Men and women who have arrived in adulthood have an image of who they should be, how they should behave, what they shouldn't do. Some men wouldn't be caught dead eating quiche! It doesn't fit their image of manhood.

People behave in ways that fit their self-image. By their late teens, young men and women don't like to be considered children. In their twenties they want to be seen as mature and grown-up. There are some things which real grown-ups don't do and don't want to be seen doing. When they were small, it was all right to go out and play, to pretend to be people they weren't, to be curious about mud and piles of sand and strange new wonders, to curl up in their parents' laps for bedtime stories, to cry and squeal with laughter, to imagine themselves fairy-tale characters, to build cities and cars and castles out of scraps.

Many people who have reached adulthood put away such childlike characteristics. In that decision, they have tucked away features of their human equipment that could make them uniquely creative. Some people let that part of themselves come out on special occasions—a picnic with old college buddies, a holiday party when they let alcohol release their inhibitions for a while, a funeral when their sadness is too strong to be held back, a critical milestone reached with much relief, an invigorating workshop that opened their thinking and feeling. But they can rapidly put a lid on the child again.

Creative Child Within

If you really want to be creative, you need to cultivate the creative child inside you. It's not that you need to slip back in time or to recreate your childhood. Your childhood might have been happy and fulfilling. It could have been miserable, abusive, and filled with real or imagined demons.

The important thing is that part of you is a child, regardless of your age. As an individual, you have been shaped by *ego states—parent, adult, and child* ego states. These are *patterns of behavior, feelings, and thoughts that began developing when you were an infant.* Your brain and senses have acted like a portable video recorder, videotaping what you witnessed in significant people around you or what you were experiencing internally and externally. Those happenings, feelings, and thoughts have been woven into three states of the mind: the parent, the adult, and the child. Patterns of these inner states bear your life imprint and are critical in how you interact with people and the world.

- The **parent** contains the values that you have accumulated over the years, the messages about what you should or shouldn't do. In this state you can be supportive and nurturing. Or you can criticize, ridicule, or put down. You can turn this caring or criticism toward yourself or those around you.
- The **adult** thrives on logic. It collects the facts in a situation, listens to the values or thoughts the parent espouses, and cocks an ear toward the concerns of the child. It tries to weigh the important factors and make balanced decisions.
- The **child** has several faces. One face is natural, reacting to the raw emotions you are experiencing—anger, fear, happiness, sexual yearning, sadness, hope, desire. A second face adapts, figuring out how to get along in social situations, discovering the appropriate thing to say or do. The third face is that of a wise, little professor—intuiting or massaging things, people, or data to create something.

It is the creative and the natural faces of the child that are especially important in the breakthrough. There are five vital characteristics that tap the imagination and release innovative energy. The child (1) helps us look at the world and "nonsense" with fresh eyes, (2) finds play expanding, (3) keeps us curious and exploring, (4) willingly dreams and fantasizes, and (5) keeps us from being emotionally sterile. These are qualities that help us find uses for a product that the rational mind can't easily conceive. Each of these characteristics will be discussed briefly.

Beginner's Mind

In his remarkable book, *The Dancing Wu Li Masters: An Overview of the New Physics*, Gary Zukav firmly declared that a toleration of nonsense is essential to imagination. It is the element that allows physicists and others interested in creative thought to rise above contradictions and puzzles of the world they are exploring. It moves them beyond being technicians and maintaining their theories to carving new avenues of thought. He concluded:

"Most [physicists] spend their professional lives thinking along well-established lines of thought. Those scientists who establish the established lines of thought, however, are those who do not fear to venture boldly into nonsense, into that which any fool could have told them is clearly not so. This is the mark of the creative mind; in fact, this is the creative process. It is characterized by a steadfast confidence that there exists a point of view from which the 'nonsense' is not nonsense at all—in fact, from which it is obvious."[7]

The child in us can tolerate nonsense, even revel in it. To allow nonsense, we have to set aside our expertise with our formulas and mind-sets. We have to put our hair-trigger judgments on hold. Zukav says we need the "beginner's mind" that can be receptive to new possibilities. This requires an empty cup, rather than a cup that is full of our speculations and opinions. The creative child holds an empty cup and shows a willingness to learn.

Einstein, Zukav claimed, had a beginner's mind. It was Einstein's confidence in nonsense that produced the theory of relativity and netted him the Nobel Prize. In the face of the prevailing and proven theory that light consisted of electromagnetic waves, Einstein postulated that light was composed of particles. He was willing to live with the inconsistency, waiting until another day to unravel the mystery. Such nonsense!

Playfulness

For the serious engineer and scientist, work is often solemn business. Play might be all right for a weekend or a vacation. Mixing work and play seems to go against the parent in us.

Your child, when allowed to kick up her heels, doesn't take daily and life events as seriously as your parent. It is your child that plays with ideas and treats nonsense as a playmate. How can a stick be a gun or a corn cob and a rag be a princess? Your child finds that metaphors and wild concoctions are stimulating to the imagination and can be used or thrown away at will. Your child lets approximations and allegories fill in the gaps or make connections. You need not worry that your child will get too carried away on flights of fancy, for your parent ego state probably has enough precision to turn the child's crude but novel approaches into marketable commodities.

Your playful child can use an endless array of new combinations. Early in your childhood you likely could try many patterns with your play things. If, for example, you were playing with blocks, you could stack them in many patterns, rearranging them at will to make a new pattern. The different angles and shapes were not taken so seriously because you could always rearrange their form.

The playful child also laughs and has fun. She uses humor as a way of looking at something from a new angle. The chuckles come from having fun, feeling good about herself, a view of the world that allows a lighter perspective amid the serious, and seeing life from fresh vantage points. Humor and laughter become a creative ally.

Both Einstein and Leonardo da Vinci cultivated the humorous perspective. They kept humor notebooks to collect the good humor they heard. Humor was an imagination stimulant they used to feed the creative child.

The playful child inside you can help you be more inventive. She keeps you loose and able to try different combinations without needing to seize just one way with a clamshell grip. She uses analogies to frolic with ideas and humor to keep you relaxed and open. She gives you time to search for the inventive while withholding judgment for a while.

Exploration

The child as explorer is a more focused kin of playfulness. A child, for instance, could be told by her mother to stay out of the makeup kit. But the colors and substances in that kit had already been observed during the mother's beautifying exercises. Curiosity and impishness caused the little girl to wait until her mother was in another part of the house. Then she discovered what fantastic patterns could be made on her body with mascara, rouge, and lipstick.

Curiosity causes you to poke your nose into something. While you might be awed by and want to play with some natural element or mechanical gadget, you want answers, descriptions, or personal knowledge. Your investigation often takes the form of a focused search or venture.

Explorers of the past have begun adventures by becoming inquisitive about lost civilizations, precious metals, jungle flora and fauna, worlds under water or high in the sky, life too small or too far away to be observed by the naked eye, special compounds, natural phenomena, or the mechanisms of energy and mechanics.

It was curiosity that caused Paul MacCready to spend time on his vacation observing and measuring the circling patterns of birds. That same spirit of exploration caused him to use those calculations of birds in flight in attempting to build a man-powered flying machine. He pursued the elusive, coveted Kremer Prize that was awaiting the person who could power man through the air over a figure-eight, mile-long course. The prize became his in 1977 when his Gossamer Condor completed the required flight. MacCready's tale is described in Kenneth Brown's interview with the award winner, along with other highly curious, adventurous souls, in Inventors at Work.[8]

Unleashing the child as explorer lets you probe for insights with curiosity, determination, and experimentation.

Fantasy

One of the most important contributions of the child to creative thinking is the ability to fantasize. Children daydream and enter a world of make-believe. They pretend they are kings, queens, pop stars, sports heroes, ma or pa, space odyssey figures. At home, in school, or on a trip, they can become absorbed in mentally picturing themselves being a specific person or in acting out a role. Children learn and create by playing "what if" games. Or trying on the behavior of someone who strikes their fancy. Or creating games, gimmicks, and gizmos to help carry out their fantasy.

Open, mental fantasies of the child allow you to use your imagination. They assist your trying out an idea without making a commitment to it. They boost your creative power because images seem to trigger your inventive wisdom.

Visualization, amplified in Chapter 2, is a handmaiden of fantasy. Mental imagery allows you to turn images around in your head to see ideas from different directions. You may add or discard at will. You might create scenarios and products in your mind's eye without the heavy expense incurred in developing many prototypes. You can use your own inner synergy and imagination to explore *what-if*'s and a host of possibilities.

Nikola Tesla claimed he tested machines by turning them on in his mind, letting them run for long periods of time. He then noted what began to wear out, the bearings first, then the wires, and so on. Robert Bailey, electrical engineer and creativity instructor, said that he has used the same mental tool. To test the weak spots before building a new creation, he would occasionally turn a device on in his mind, thus saving much cost, time, and frustration.[9]

Many people, however, shut down their dreams and fantasies shortly after starting school. Their dreams became lackluster and black-and-white, their fantasies dull. The rational side of them tamed down or blotted out fantasy because it seemed too nonsensical or whimsical. Another part of them might have felt that imaginary images were to be feared or would cause them to go out of control. However, when we turn and face the "monster" chasing us in dream and fantasy life, often we discover that the pursuing ogre diminishes in size and scariness. What often appears, when we let the dream or fantasy unfold, are creative solutions. This can be true whether we are dealing with sleep or work time dreams.

An enriched fantasy life will allow you to take full benefit of many creative thinking techniques. Imagination requires fantasy.

Emotional Coloring

In laboratories and fields, in plants and testing rooms, data must be carefully observed, measured, scrutinized, analyzed, and documented. Crystal-clear logic is needed.

Star Trek's Spock, the half-human, half-robot android is a metaphor for the highly analytical approaches called for in many engineering and scientific tasks today. Spock's laser-sharp mind was often on center stage, with his muffled human emotions sometimes entering the picture to puzzle his rational mind. Spock could get away with such a fine-edged show of rationality in the twenty-third century. The scripts called for it.

Twentieth century researchers and developers, however, need to apply both the rational and arational mind to the problems and opportunities of today. The natural child is needed, but not just to balance the rational.

Feelings help fire up the imagination. Emotions trigger memories and associations that tap your past and form connections between experiences and solutions. Feelings enrich fantasies, color dreams, excite play, arouse interest, enliven energy, focus direction. A spate of emotions can add heart to discipline, humorous sparkle to deadly serious endeavors, healthy vulnerability to the uptight mind, empathy to another's idea. Selected sounds, sights, smells, touch, and taste can stimulate creative thoughts by adding emotional coloring and stirring inventive associations.

A bland mind produces lukewarm ideas. *Eureka*, serendipity, and synergy require a mind that is emotionally as well as intellectually fertile. Certain emotions, of course, could throw off a creative effort. Approaching a problem or opportunity with a halo of "the blues" or excessive anger or manic mood or deep distrust could dampen, distort, or discount the use of imagination.

A thinker insensitive to his colleagues' feelings might not help his group's attempts to produce a new and innovative product. An introvert intimidated by a dominant team member might feel mentally inhibited and not contribute her off-the-wall but valuable insights.

To use your natural child in creative thinking, you have to see emotions—the exciting and the cautious—as genuine, healthy, and spontaneous. You have to cultivate, not avoid, their sensations and moods. Once you do, you will have a more direct and colorful route to your imaginative wisdom.

POSTSCRIPT

Breakthrough energy doesn't come from just soft focus or just childlikeness. These characteristics need to work in concert. When they perform in harmony, synergy is free to do its wonders inside the person in the same way group cooperation produces exciting new thoughts.

Breakthroughs await discovery. Sometimes we just stumble onto them by chance. Or we nudge ourselves in their direction by using a mental guide. Flights of the imagination uncover possibilities where none were seen before.

4 Breakthrough Discovery Process

Don't refuse to go on an occasional wild goose chase. That's what wild geese are for.
—ANONYMOUS

The search for truth is in one way hard and in another way easy, for it is evident that no one can master it fully nor miss it wholly.
—ARISTOTLE

We dance around the ring and suppose, but the secret sits in the middle and knows.
—ROBERT FROST

Creative ideas may pop into your head from out of the blue. There may seem to be a magical quality to your thinking. But usually new options that appear in your imagination bubble to the surface as the result of mental processes that are hidden to you. When you are confronted with a situation that requires a new approach, those internal processes keep churning along, seeking solutions. Even during sleep or while taking a break, your mental energy is directed toward the problem or its solution. "Spontaneous" or "breakthrough" ideas often come after lots of hard work analyzing and mulling over a problem or opportunity, often while doing something disconnected to working on the problem.

Creative thinking need not be haphazard. Hit-or-miss methods, of course, can help you find solutions. As Figure 1 shows, with such approaches you usually identify or use only a small portion of the range of solutions available to you. Often hit-or-miss ideas are of lower quality. But, if you are looking for a solu-

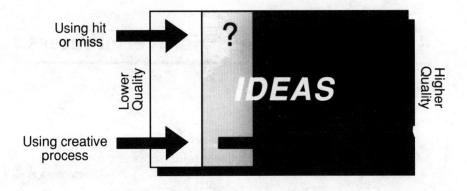

Figure 1. Chances of reaching more imaginative solutions.

tion that will be a Band-Aid or an adaptation of existing solutions, breakthrough ideas don't matter. However, your chance of reaching more elegant, useful, or novel ideas increases significantly when you use creative processes.

You can intentionally induce the flow of imaginative options by knowing and using the creative cycle. This cycle is a series of phases you can follow, which will increase your chances for getting more, and higher quality, ideas. When these phases and their powerful processes are followed and allowed to work, you release amazing creative potential and energy. The steps in this cycle provide the catalyst for those special qualities of the human instrument—soft focus and the creative child—to ignite further imaginative power.

THE FOUR PHASES

Four primary phases comprise the creative cycle, imaginative processes for individuals or teams. Within each phase are steps, decisions to be made and processes or techniques to follow. The creative cycle calls for discipline and use of both rational and arational mental approaches. The emphasis in this book is on teaching effective ways of generating, capturing, and pinpointing inventive thoughts. Other writers describe analytical, decision-making, or legal aspects of the creation process. Here's a terse summary of what the phases entail:

- **Determine the target:** Gather data about the situation. Analyze the factors. Take aim on a part of the situation that could use some fresh ideas. Settle on any needed boundary conditions—specific requirements or desires. Shape a question as a prompt to guide your search for ideas.
- **Search for options:** Disrupt common perceptions. Follow a creative thinking strategy. Generate ideas.
- **Check for fit:** Match the new ideas against the requirements. Choose the best. Determine implementation support and roadblocks.
- **Take action:** Set the idea in motion. Build flesh around it. Design it. Model it. Test it. Simulate it. Produce it. Market it.

These phases are a cycle of processes (see Figure 2). They revolve because solving a problem or identifying an opportunity may require a series of iterations and refinements. As you get new information or look at something from a different perspective, you may need to jump forward or backward in the cycle.

Most people who pick up this book will not be novices to problem solving. Since you were knee-high to a grasshopper, you have been figuring out how to solve problems. How to reach a rattle just beyond your grasp? How to tie your shoes? How to put the clock that you knocked on the floor back together? How to get someone to like you? How to pass a test at school? How to handle a new job assignment that stretches your abilities? As you tackle the types of problems just described, you may sometimes intuitively know the answer. At other times you may gather piles of facts, lay them out on the table, and put the factors together like a jigsaw puzzle.

The creative cycle is a process for getting good ideas consistently, when the problem or opportunity is too important to be left to just "winging it" by hit-or-miss methods. Figure 3 lays out the basic flow of the creative thinking process.

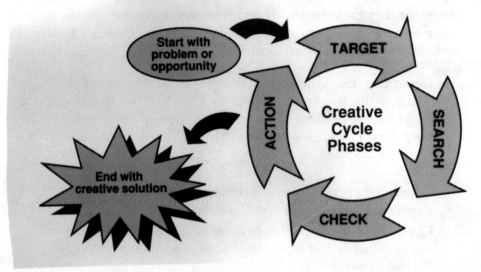

Figure 2. The creative cycle.

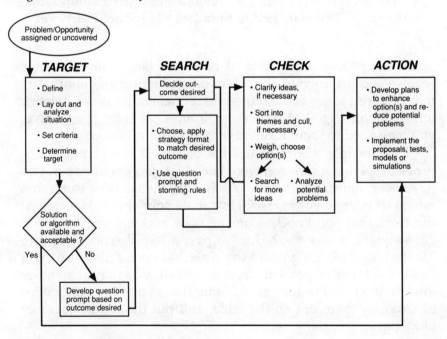

Figure 3. Creative cycle process flow.

LOOKING AHEAD

We'll take each phase of the creative cycle in turn in Chapters 5, 6, 7, and 8. This creative thinking process is colored with the abilities of the whole mind. In addition to the use of logic and a rational sequence, the arational—intuition, big-picture thought, the nonverbal—will be woven into each step. The arational hue provides the breakthrough punch. The search phase is where an explosion of imaginative thought is sought. However, fresh, intuitive, and arational perspectives can spark creativity throughout the cycle.

The breakthrough qualities of the creative cycle rest on having the right mental attitude and approach, as well as the kind of climate in which to do creative thinking. We have already looked at the special qualities of thinking that are needed. Most of the climate-setting will be tackled in Chapter 11.

One vital aspect of the environment, team communication, will be described as we move through the steps of the cycle. Team communication will make or break the creative thinking process when it is used in a group. Tips for effective interaction will be provided at the end of each chapter on the four phases.

5 The Target: Taking Aim

> *The uncreative mind can spot wrong answers,
> but it takes a very creative mind to spot wrong
> questions.* —ANTHONY JAY

Great inventions of the past arose to meet a need. Early man threw spears at wild animals in an effort to provide food. But the shifty creatures could often dodge and move faster than the hunter. Whoever invented the bow wanted to diminish the human disadvantage and put meat on the table more efficiently.

As humans progressed, their eyes grew bigger than their backs. They kept seeing heavy objects they wanted to lift without straining their bodies. Someone—either ingenious, lazy, or both—decided to do something about it and created a highly useful labor-saving device, the lever. Later came the pulley.

Human predicaments were turned into more favorable conditions. Invention of the bow and the pulley arose from an understanding of what was and what was desired. Sound and original thoughts have emerged to meet a need, to help get us from a current state to a desired state.

The first step of the creative thinking process, *the target phase, is to get a good handle on a current situation and on a clear image of the condition we'd rather have. We then zero in on the arena where we want to make a change.* Actually generating proposals to make a change is the task we undertake in the next phase, the search.

In the early 1970s, I encountered a useful problem solving format then making the rounds in a number of workshops and organization effectiveness efforts in the Northwest. Developed by Frederick Fosmire and John Wallen, the elegant process, called

STP (Situation-Target-Proposal), was a three-pronged way of arranging data for problem solving. Many other problem solving formats have been developed over the years. I discovered in countless creative problem solving sessions and settings that the STP approach is a compact aid, a fine mental template for following the essential features of problem solving, without getting too distracted or bogged down in details.

In the STP model, interaction flows back and forth among the three interconnected elements. Knowledge about any one element implies information about the other two. Aspects of Fosmire and Wallen's situation and target steps have been blended into the target phase presented here.

SITUATION–TARGET INTERACTION

The target phase, which lays the groundwork for directing a search for ideas, has two primary parts: (1) a current condition or state of the art, and (2) a target or expression of a desired, hypothesized, or ideal state.

- *The current condition or state of the art is composed of influencing factors, helping and hindering forces, and causes and consequences.* To uncover the current situation, you pinpoint variables that affect or might affect the present. Your descriptions should include gathered facts, stated opinions, and gut feelings. Your compilation of these variables may characterize the level of knowledge and technology known or used. A major task in investigating the situation is to lay it out in such a way that you can get to the heart of a problem or opportunity. You want to be able to peel away the extraneous and work with the essential factors.
- *The target or expression of a desired, hypothesized, or ideal state is the condition you'd like to have.* Aims, outcomes, intentions, goals, hypotheses, and purposes are forms of targets. Your task is to develop statements that pose the key future circumstances, a state without conflicting or unwanted features.

Discovering the current and desired state can be as simple as learning the plot in a comic book or as unnerving as unraveling an Edgar Allen Poe mystery. In some circumstances, you can home in quickly. At other times, you are required to engage in a complex investigation, like that undertaken by C. Auguste Dupin in "Murders in the Rue Morgue," Poe's invention of the detective story.

It's Paris in the mid-1800s. Madame L'Espanaye and her daughter die violently at night amid mysterious circumstances. They are the sole occupants of a large stone mansion, the Rue Morgue, guarded by a strong fence and iron gate. With wild shrieks coming from a fourth floor apartment, neighbors gather at the gate, trying to get in. Once two gendarmes and the small party of neighbors break open the gate, they rush into the house and begin searching it. By the time they reach the second floor the cries end. But those noises are replaced with a burst of strange voices, one speaking in French, the other voice not identifiable.

Once on the fourth floor, the searchers break into a back chamber, which is locked from the inside. The room is a shambles, with the bedstead and other furniture broken and scattered about, a bloodstained razor on a chair, coins scattered on the floor. The marred body of the daughter is found stuffed up the chimney. After some time the mother's gruesome body is found in a small backyard behind the house.

The authorities search the house thoroughly, combing it for clues. They are unable to discover who is involved in the murders or the route of escape. Dupin, a retired, former detective, joins the hunt. Applying his masterful mixture of analysis and imagination, he gradually puts the puzzle together.

To solve the mystery, the detective continually moved back and forth between his discoveries of the factors in the murder situation and the target or desired state. He started from a general, hoped-for outcome—solving the crime—and began to alter the target as he learned more about the situation. Using the current and desired state elements of the target phase, Dupin, as he looked deeper into the problem, might have broken the data down something like this:[1]

Current state:	Desired state:
Initial assessment: Two gruesome murders. No motive apparent.	Killer in custody.
Intermediate assessment: One woman crammed up a chimney, another torn body in back of the mansion.	Killer, capable of enormous feats of strength, apprehended.
Final assessment: A window, which the police thought couldn't be opened, was found to be capable of being entered. A lightning rod ran up the building close to the window. Imprints on the neck of the daughter didn't match that of a human hand. A ribbon with an unusual knot was found on the ground close to the window.	Agile human capable of tying unusual knots, accompanied by a nonhuman creature—capable of great agility, power, ferociousness, as well as speaking a strange gibberish—located.

With the array of clues filling in the puzzle of the current state, the desired state changed and became clearer. Dupin laid a trap for a sailor, someone who could quickly climb a lightning rod, tie that unusual knot in a neck ribbon, and could have come in contact with an ape on his travels. The trap worked. Dupin found the culprit and the reason for the murders, as Edgar Allen Poe wove a yarn of mystery and of a detective with both imaginative and logical reasoning.

As detective Dupin showed, there is creative tension between the current situation and the result or outcome you'd like to have. Foresight and genuine problem understanding—as well as probing, alert, and imaginative exploration—are often required to see what is present and to define what is needed.

You can't have your arms around a problem or opportunity without bringing the two parts into focus. One, the situation, may be messy, a jumble of feelings and factors. The other, the target, seeks clarity by sharpening the vision of what should be. To isolate the real problem, you often have to work back and forth between the situation and the target until a clear picture of the real problem or opportunity emerges.

To get that snapshot you could take a logical, sequential approach to unravel both of them. If you were to take a non-linear, wholistic, intuitive perspective, however, you would add insights to a bigger picture of the scope of the issues and desired outcome.

ZOOMING IN OR OUT OF THE SITUATION AND TARGET

There is something you need to do before you plunge into the creative cycle.

Grab a cup of coffee. Sit down and relax a bit. Reflect on how close you want to get to defining the problem and to sizing up a solution. If you declare that you want to get the thing solved now so you can move on about your work, you could be right on the money. Or you could be fooling yourself by going for the quick payoff. You may be bypassing a long-term gain for a short-term resolution—like not focusing on saving for a rainy day or an exotic vacation but blowing your paycheck on present thrills and spills. Both may be needed, but at the appropriate time.

Pull out your mental zoom lens and determine the distance from the situation and target that will give you the most mileage for the solutions you develop.

If, for example, you want to explore a piece of land, you could do so from different vantage points. You could crawl on your hands and knees—parting blades of grass to see what is hidden under them, turning over a rock to discover the texture of soil or creepy things living under the surface, feeling the texture of living and inanimate objects. That would give you a very intimate knowledge of that land parcel, one that would be different from walking about, which would give you a wider perspective.

Climbing high into a tree to view the same area, you would lose some of the detail but gain a wider vista. Perhaps you would see the deer silently watching from another clump of trees or notice how lush the vegetation was around the gently flowing stream. Hovering over that territory from a helicopter would give you the greatest breadth. New relationships of the land would come into view: slopes and contours, multiple colors and plant life.

Your mind accomplishes the same basic feat, zooming in and out on a subject to study it from different viewpoints. In the target phase, you can use both mental and process zooming. You can determine the distance from which you would like to study the situation and set targets. Apply this idea of zooming to targeting a problem.

Imagine that you are employed in a company that manufactures various electronic products for home entertainment. One of the parts on the turntables you manufacture has been giving you fits, inconsistently meeting your quality specifications. The culprit is the relay, the mechanical tripping device that turns sound on when the stylus makes contact with a record.

You determine that the problem is the placement and adjustment of the relay. You find that, from turntable to turntable, there is a little difference in where the device is attached, causing inconsistent tripping of the sound mechanism. Currently the relay is attached to the turntable by hand.

Apply your mental zooming. By adjusting your focus on the problem, you can arrive at different targets and, thus, varying solutions. Most organizations zoom in close to solve the immediate problem. With a close zoom, here's what your problem solving might show:

- **Target:** *Improved manual installation.*
- **Possible Solutions:** *Provide retraining. Have an industrial psychologist identify who has the greatest aptitude for accuracy on the task, then use only the most accurate employ-*

ees. Emphasize quality at employee meetings. Develop a template to guide placement of the relay.

Zoom out a little way. Take a little broader perspective, where you might be willing to change your processes or methods. At this distance, your problem solving might look like this:

- **Target:** *New, error-free installation.*
- **Possible Solutions:** *Automate. Rebuild the turntable to allow molded jacket into which the relay could be placed. Tune the relay after it has been installed, recycling rejects.*

Be bolder. Zoom some distance from the specifics of the problem. If you were to stand back and not be locked into that mechanical relay, you might be able to discover new ways of triggering sound that would improve efficiency, quality, and durability. A problem solving session might then produce the following:

- **Target:** Mechanism for turning on sound that can improve responsiveness, effectiveness, and quality.
- **Possible Solutions:** Optical laser. Electromagnetics. Acoustics. Infrared light. Sound signals. Pressure on the needle.

Zooming allows you to make decisions about where you want to put your focus—on the short or long term, on better twists or innovations. While you may solve the immediate problem by zooming to close range, you may also just produce a Band-Aid. In the above example, for instance, effort put into retraining employees for greater accuracy may still give you spotty quality.

If you want a solution that could give you greater returns and higher quality, despite potential retooling, try zooming to a broader perspective. You may make big gains in the long run. By standing back to view the problem, you have the opportunity of looking at a much wider array of possible solutions, some that might cause you to stretch for new technology and more novel approaches.

TARGETING STEPS

You have looked at the two critical segments of the target phase—the situation or current state and the target or desired state. You have also observed, in "Murders in the Rue Morgue," how the target can change as you gather new situational data. And you have seen how applying a mental zoom lens to a problem or opportunity can give you different targets that lead to varying idea suggestions.

Now plunge into the steps outlined in Figure 1 that will carry you, in a disciplined way, through the target phase. First, follow the steps to help you understand what information could help you zero in on the current and the desired conditions. The degree of effort to put into each element depends on the complexity and nature of the problem you are trying to solve. Studying the case of the parched land, described later in this chapter, will help you understand how to apply the targeting steps to a specific problem. Here are the steps:

1. Define the Concern

> **Build the definition of the issue. Recognize, however, that the first attempt may be crude and just serve as an initial cut. Make a statement summarizing the crux of the problem or opportunity that includes a brief, even terse, description of the situation and its impacts. Keep the definition free from technological jargon. Include, if known at this point, what features of the situation are wanted or unwanted.**

If your definition doesn't fit after gaining a fuller understanding of the situation, refine it. If the statement gets you started on a survey of the current condition, it has served its purpose in focusing your attention and your next efforts. Don't try to have a perfectly worded statement at this time.

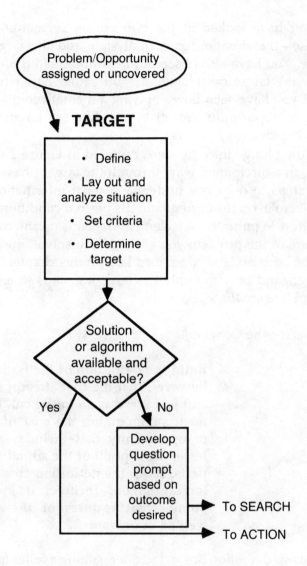

Figure 1. Target process.

Here is where one's psychological type, described in Chapter 2, intrudes. Remember those sensing-type individuals who prefer facts, being realistic, and dealing with the present? They will approach definitions, standards, and requirements from a practical standpoint.

Intuitive types, on the other hand, often add their own big-picture perspectives regardless of the statement that is written. Even while the definition is being composed, the intuitives may be looking beyond the constraints of the present situation to the forces that set the boundaries. They may be glancing ahead in their minds to how a broader context needs to be considered, naturally zooming out to get an overview snapshot. Make sure both sensors and intuitives have input into the definition.

2. Lay Out and Analyze the Situation

> **Gather facts, impressions, and opinions to describe the forces, factors, causes, and consequences in the current problematic or opportunity-laden situation. As much as possible, determine the elements at the core—conflicts, unwanted features, deviations, inconsistencies in technical or institutional systems. Use an appropriate analysis method.** (A variety of analysis techniques is presented following the description of the basic targeting process steps.)

You can gather data from various sources—your own knowledge, colleagues, plant operators, lab technicians, other disciplines, literature, customers, project sponsors, theory, related fields, and so on.

When you are investigating the situation, the amount of detail you lay out will depend on the type of problem under consideration, the specific area you want to explore, how broad a picture is necessary, and the analysis method you choose.

3. Set the Criteria or Boundary Conditions

> List specific requirements if there are "must have" constraints.

Here's another time that your psychological type matters. People who have a practical, sensing, "quick closure" bent often like to have the constraints up front. They want to detail criteria before they come up with ideas. They don't like to generate options that might be wasted or not meet the requirements.

Intuitive types often chafe at this early detailing of the boundaries. Many times they believe that setting the criteria before searching for ideas is putting a noose around their thinking. The challenge to them often comes from generating ideas that can get around or obliterate what seems to them like unreasonable or unrealistic constraints.

This step is put in the targeting process here because sometimes major givens restrain the development of a solution to a problem. Such constraints must be understood, sometimes as a limitation and other times as challenge.

If, for example, the person who is going to be making the final decision on your recommendation has declared that he would fire anyone who comes to him with a solution costing over $50,000, you are dealing with a requirement regarding your end product. That constraint should be acknowledged as part of the analysis, then noted as a criterion here.

If you are searching for a drug that can attack a virus hiding in the brain, you have to consider the blood-brain barrier an important constraint. This biological wall is set up to allow nutrients to get to the brain while protecting it from potentially harmful substances. You might consider the blood-brain barrier an unchangeable given. Or you could list this barrier as something a solution must resolve, while also accepting it as a challenge. You then might try a novel solution to break through or to fool the system.

Potential advances in treating difficult viral diseases may come, if current promising research stays its course, from refusing to accept the blood-brain barrier as an unsurmountable dividing line.

Breakthroughs have often come from sliding around, knocking down, or finding a way through the maze of a boundary condition.

Based on the issues, systems, and processes you are facing, you may or may not set limiting, bounding criteria. The dilemma is to recognize and acknowledge the realities or givens of the situation without constraining your ability to come up with boundary-violating ideas. Many constraints have been found not to be fixed walls. You might prefer to create criteria in the check phase of the creative cycle, when you need to evaluate the ideas you have generated.

4. Determine the Target

Take aim. Shape the target by deciding where to focus your idea-generation efforts. Make brief statements of what you'd like to see in your desired state. You could describe roles, outcomes, activities, or ideal technical or physical circumstances. Make the statements as if what you want were to happen now. (Remember the examples of targets in the discussion of "Murders in the Rue Morgue" and of zooming.)

Steer away from solution suggestions. If you uncover options at this step, jot those ideas down for use during the search phase.

You may be tempted to make solution statements in this phase that describe a need: "Need a hyperspeed trim reducer." "Need a Hovercraft to distribute beer at soccer games." "Need a user-friendly thermal enhancer." Statements of need suggest conditions not being met in the current state, but are posed as answers. Don't get trapped in coming up with solutions at this time, but don't toss out the ideas either. Save them for the search phase.

You may also want to rephrase those suggestions to ask what would be happening in the desired state if your real need were met. The "need for Hovercraft at soccer games" may suggest a target of "rapid distribution of beer supplies to meet demand levels during soccer games."

When brief target descriptions are stated as if they were actually occurring, you get a clearer picture of what the future condition could look like. This helps you design the possible ways to arrive at the desired future.

5. Determine If a Solution or Algorithm Is Available and Acceptable

> Occasionally you may come across a situation where an answer is obvious. A sensible solution comes to you, either through an intuitive flash or by seeing where you can apply a tried-and-true fix.
>
> If you decide that you are satisfied with a solution that has come to you thus far, you proceed to implement it. However, it would be well to go to the check phase to assure yourself that you have thought through the potential problems inherent in your solution. Then you move to the action phase to add appropriate planning and steps for setting your option(s) in motion.
>
> Before rushing the idea to the lab to test it or to the computer to simulate it or to the drafting table to design it, do two things: First, sketch it to create a visual image of your solution, if that is possible. Second, do a patent or literature search to find if the solution has already been developed or tested somewhere.
>
> If your analysis of the situation hasn't turned up a sure-fire solution, go to step 6.

An obvious solution, formula, procedure, activity, or answer may have popped into your mind as you developed the target. The mind often works in multidimensional forms, sensing solutions before all the data are out on the table. Sometimes in the interplay between determining the current and the desired state, gaps are noted and potential solutions appear. The early ideas tend to be more commonplace, more practical. Again, jot them down so they don't get lost.

Some problems will require the breakthrough. Yet don't forget the saying of the wag, "Great minds travel in the same circles." Before laying out the money and time to pursue development of a new process, gadget, or theory, you may make many strides by doing some investigation first. A wave of thought and research around the globe has preceded many discoveries—insulin, computers, manned flight, the black hole, molecular mysteries. Your great idea may be just that.

To use a surfing analogy: someone else might have caught the big wave just before you and reached shore first. You should capitalize on a prior technique, on what other discoverers have learned. Then your refinements of someone's development may help you ride the crest of the next wave. Alternatively, if you desire unique solutions, you avoid needlessly reinventing the wheel.

Of course, there are many of you who have an insatiable desire to do it your way. Looking at what someone else has done inhibits you. R. B. Benjamin contends that, in some circumstances:

"I often prefer to attack a problem without knowing the previous patent or commercial art, and thus avoid following beaten paths.... I prefer to have my mind free for any ideas that may come to me without being burdened with visions of past performances."[2]

Likewise, while you may not want to reinvent something, you want the freedom of not being bound or swayed by old modes and mind-sets.

To produce a breakthrough, you may have to look over your shoulder at other great ideas while also rummaging through your own imagination.

6. Develop a Question Prompt

Form a question that will prompt your thinking as you search for ideas to change the cur-

rent situation to the condition you want. Follow these rules for the question prompt: (1) Start with *what* or *how*. (2) Don't imply a solution. Give yourself room to explore a breadth of possibilities. (3) Don't seek more analysis or situational data. The initial data collection should have been concluded. (4) Include a desired state you want to satisfy.

By taking this step, you are declaring you will commit the energy, resources, and frame of mind to engage in generating proposals for satisfying a desired target. Step 6 lets you shape a question that will build a bridge between targets and the search phase.

Your target statements provide clues for what to put in the question. Most target descriptions, in fact, could be turned into a prompt. From two target statements in the problem of inconsistent relay performance in turntables, here are a couple of prompts that could help you launch a search for solutions:

- How can we improve manual installation of the relays?
- What are new mechanisms for turning on sound that could improve auditory responsiveness, effectiveness, and quality?

The first question is narrower in scope and would hem you into generating ideas pertaining to your current relay and manual installation. The second one provides for a wider range of possibilities and would allow you to come up with ideas to satisfy a wider range of possibilities. If you want more freedom in your search, you'll go with the second question prompt.

Now that we have previewed the steps in the target phase, let's explore an example of their use. Otherwise the target phase may seem as elusive and cumbersome as trying to eat Jell-O with a toothpick.

The situation to be explored, the case of the parched land, was suggested by a participant in one of my workshops. It concerns the shortage of water in the Southwest, but has applications throughout the nation and the world. This problem is broad, requiring zooming out to include many technical, political, and environmental facets.

CASE OF THE PARCHED LAND

We'll use a mythical southwestern state, DesertSun, as the focal point for the case of the parched land. This case is ripe with social, political, economic, and environmental, as well as technical, considerations.

Setting

DesertSun is a beautiful, sun-drenched state with a warm, dry climate attractive to a growing number of citizens and corporations. The state has scattered waterways fed by runoff from mountain snow, no major lakes, and few aquifers of consequence. The limited natural supply of water is being depleted by the spread of irrigated crops, growth of industry, and a housing boom. Surrounding states believe they have no surplus water. The governor has set up a commission—including representatives of industry, academia, and the technical community—to examine the situation and to make recommendations for the state.

Concern

With the general sense of DesertSun's dry climate and strained natural supply of water, we can make a first-cut definition: DesertSun's water supply is insufficient to meet its growing long-term farming, business, and homeowner needs.

Analysis

For the case of the parched land, a number of methods could be used to systematically gather and display the situation for analysis. Since this is a broad, multifaceted problem, one fruitful approach would be to develop a list of factors that influence the current condition.

Granted, this is too big a problem for us to fully explore here.

But for illustration purposes, I want to show the rudiments of an analysis method of the target phase. I don't want to get sidetracked in technical detail. When you use the process, you need to add the specific data important to your situation. Here's a partial, fairly nontechnical listing of influencing factors:

- Arid conditions stretch over most of DesertSun. Much former desert has been turned into productive farmland, adding a strain to the water supply.
- Abundant sunshine, dry air, and beautiful scenery has lured many corporations, retirees, and workers. The influx has resulted in a growing demand for water.
- Rainfall is low. High temperatures in the area speed evaporation. Although rivers create a web over the state, the water supply is inadequate to meet the growing demand. In addition, the recent rising temperatures, conjectured to be caused by the reduction of the earth's ozone layer, have put added pressure on the delicate water supply.
- There are some current attempts to conserve water by curtailing car washing and lawn watering in critically dry periods and by rationing the water supply to farms and industries.
- Attempts to obtain water from surrounding states have been met with controversy and little progress. One major study contends that any water received from surrounding states would damage the water supply in the other states and not adequately meet DesertSun's needs.
- Primary uses for water are quenching thirst; growing plants and food; cleaning humans, cars, homes, etc.; providing power for home, farm, and industrial use; disposing of wastes.
- Humans know water is essential for life and are interested in beauty, shade, cleanliness, and growing things.
- Water can be in vapor, solid, and liquid forms.
- Chemical pollution concentrations—and their effects on human health—tend to increase rapidly as water reservoirs fall during drought conditions.

Certainly there are more factors that have to be considered. This is just a beginning. Additional data could be obtained to amplify and quantify each element above. We might, for instance, want to obtain projections of growth or anticipate special circumstances that could deplete the water supply.

Criteria/Boundary Conditions

The governor of DesertSun has been getting increasing pressure from industrialists, homeowners, and environmentalists to take some action. In his initial charge to members of the commission, he suggested that the following guidelines be reflected in the recommendations:

- *Reach consensus as much as possible.*
- *Use bold thinking.*
- *Strive for long-term solutions.*

Target

Here are a few phrases to describe what would be happening in our desired situation:

- DesertSun homeowners actively conserving water.
- Mechanisms reducing the normal expenditure of water.
- Water more available for home, farm, and industry use.
- Substitutes developed to replace human cravings, desires, and uses of water.

Of course, there could be more targets. You might think of some to add to the list.

Solution Availability

While developing the steps to this point, we are likely to have thought of some solutions—penalties for watering lawns, bricks for the water closet, water-saving shower nozzles, more efficient irrigation, recycling of industrial water. Those ideas have possibilities for getting us from the scarce water supply situation to the desired state targets. Further information could be sought to amplify the potential of these spontaneous, well-known solutions.

Since there are many dimensions to this case, a basic need is to look for the most promising solution areas. Many ideas need to be put on the table so we can see the range of possibilities that could be tackled. Techniques will be needed to uncover the scope of good ideas, then to home in on the most promising.

Question Prompt

Here are two:

- How could more water be conserved?
- What are ways to increase the availability of water for home, farm, and industry?

The first one is narrower in scope and would hem us into generating ideas that have to do with conservation. The second one provides for a wider range of possibilities and would allow us to come up with ideas to satisfy all of the target statements. If we want more freedom in our search, we'll go with the second question prompt. Our next stop with this case will be in the search phase.

Before we go to the search phase, though, we'll examine analysis methods that can be used in the target phase, see how we can add an arational touch to the targeting process, and sketch communication tips that could enhance this first phase.

ANALYSIS METHODS

What do these folks have in common—a computer whiz building the next generation of supercomputers, a bioengineer fashioning a robotic arm operated by nerve impulses, and a behavioral scientist assisting a company's attempts to instill creativity in its people? They all need to corral the essential variables that play on their particular tasks, to prevent being overwhelmed by the multiple and detailed factors that bear on a mechanism, function, system, or activity. And the more radical the change they want to make, the more the variables and possibilities that have to be considered.

Corralling and analyzing the variables in a situation can be as easy as sitting in your rocking chair and mulling over the factors of a simple problem. Or it could require numbing hours in front of a computer to sort, display, model, weigh, and compute a multifaceted situation. The right tool is needed for each task. You wouldn't use a fishnet designed for hauling in halibut to capture minnows, or a fly swatter to kill a darting bat. You need to choose the analysis tool that will get the job done.

You may have a favorite method for gathering and analyzing data. Your favorite might be building a decision tree on a chalk board—probing a situation with questions, laying out the branching factors from the general to more specific. Or, to get concrete data, you might like tracing the current state using a new software program or the latest statistical method. Your favorite could be gathering a cross section of colleagues for a morning group-analysis session.

I'm suggesting a variety of effective analysis techniques that can be used during the target phase, methods that could complement the tools you already use. These are mainly nonstatistical tools you could use as an individual or while working with a team.

Some of the methods presented here can also be used during the check phase when you want to review newly generated options for potential problems. Feel free to use these approaches separately or combine them in any manner that enhances your creative processes. Other books, devoted entirely to various forms of analysis, describe analytical tools in greater detail.

Determine the kind of analysis and zooming distance wanted. Do you need an overview? Zoom back to get the big picture, analyzing for breadth rather than detailing many variables. Do you need an in-depth evaluation of an element or part of the whole? Zoom in to examine particulars. Are you after an examination of relationships between many elements, a structured picture? Zoom out so you can view system interrelationships. See Figure 2 for a schematic contrasting the level of zooming, the analytical picture, and type of detail required. Choose analysis methods to fit what you need.

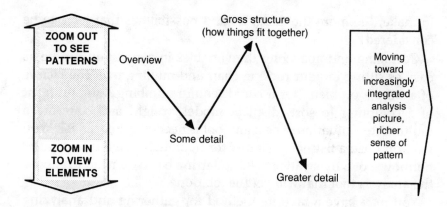

Figure 2. Relationship between analysis breadth and detail.

There are many helpful approaches to analyzing data. To be useful an analysis method should (1) help you collect data, (2) display it so you can study relationships between elements, and (3) allow you to zero in on the most critical factors.

Here are several approaches that can be used quickly on a simple problem or can be expanded to include complex issues with numerous, interacting factors. They can be linked to help you dig into different types of material or meet different needs. Some situations are so complex that an analysis has to be done on several parts of a larger issue. It may take a long time to complete an investigation of a complicated problem, such as the DesertSun case.

Situation Analysis

Situations include such things as events, forces, processes, methods, elements, facts, opinions, myths, behavior, attitudes, and expectations. While some of these are central and others may turn out to be peripheral to your problem, all can add perspective to a portrait of the current state. The picture you create of the situation is often needed to determine the scope and shape a target should take.

1. Identify situational characteristics. List features, influencing factors, special concerns, and/or assumptions.

2. Define key deviations from a goal, norm, or course. What specifically in the situation departs from expectations? What is the location, uniqueness, size, or timing of any deviations? Gather information. Test for clarity of factors if necessary.
3. Identify who is involved—the group or individuals who could be affected by decisions or solutions; sponsors, clients, or customers who have to be satisfied; and possible champions. Briefly describe their contributions.
4. Use methods below to pinpoint other aspects of the situation.

A very detailed situation analysis, for instance, was required in the aftermath of the *Challenger* disaster.

A Presidential Commission carefully, methodically laid out the characteristics in the situation surrounding the accident. Special features were the launch pad, main engines, fuel tank, rocket boosters, solid rocket motors, the orbiter with its equipment, and the interface between the payload and orbiter.

Hypotheses about faults or problems in each of those areas were conjectured and tested. Data from the mission, the wreckage, tests during the building of the spacecraft and of specific systems following the disaster, and design issues were examined to explore factors that could have led to the accident. Design data, photographs, computer simulations, and a second-by-second analysis of major events provided some of the evidence needed to start narrowing down major potential causes. Key individuals were quizzed. Hard questions were asked to scour information for the location, uniqueness, and timing of departures from what was expected.

Once the O-ring became suspect, it became the culprit of choice only after much careful weighing of such contributing factors as the weather and potential vulnerability of the joints in the right solid rocket booster.[3]

Situation analysis can be applied to a highly complex, dramatic concern, such as the space shuttle accident. It can also be used to explore the current conditions surrounding an organization's reward system for its professional staff or a problem in the lab or on the assembly line.

98 CREATIVE THINKING AND PROBLEM SOLVING

Often situation analysis is used to ferret out the causes that most affect the current condition, as was done in examining the *Challenger* incident. This method helps wade through a jumble of facts, forces, and opinions to arrive at the critical causal factors that require more attention. In your study, you develop target statements to pinpoint conditions you want to highlight or change. Then you look for solutions by generating proposals in the search phase. Usually the crucial causes are only a part of the whole situation, as depicted in Figure 3. Unlocking the puzzle takes a helpful process, smart thinking, keen observation, and good judgment. A clearly defined problem—which includes both the current and the desired state—is needed before you can search for a creative solution.

In summary, through situation analysis you make sure the key factors are on the table, then evaluated. Deviations from your

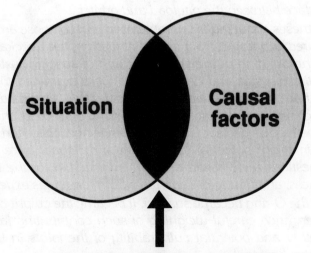

Figure 3. Relationship between the situation, causal factors, and prime causes.

goals or expectations are examined. Causes and consequences are sought, with impacts highlighted.

Force Field Analysis

This method is based on Kurt Lewin's theory that situations, events, processes, products, people, and goals operate in and are maintained by an array of forces of varying impacts. An analysis of the forces helps pinpoint where adding, reducing, or eliminating a force can move the targeted area closer to the desired result. Movement comes not just by piling up more positive forces but also by reducing those negative forces in the way. Adding only helping forces often builds resistance, because the negative forces remain unchanged.

1. Zero in on forces surrounding a specific aspect of the situation. Suggest a goal you'd like to meet to move you in a positive direction. Suppose, for example, you said you wanted to affect how homeowners in the state of DesertSun conserve water. Write a brief target statement at the top of a page, such as, "Re: Increased homeowner water conservation in DesertSun."
2. Let a vertical line on a page or newsprint represent the status quo or current equilibrium of forces. (See Figure 4.)

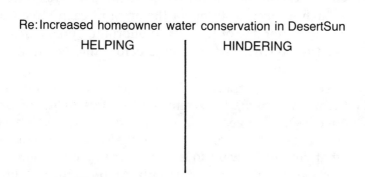

Figure 4. Force field analysis example, step 2.

3. Identify forces helping or hindering the attainment of a goal, direction, or satisfactory result. Include facts and perceptions. Collect plus and minus forces as you think of them, for thoughts about a positive force may trigger ideas of negative pressures. One aspect of a situation could even be on both sides of the line—part constructive, part inhibiting. Note Figure 5.

   The first cut calls for a free flow of information, in storming fashion. Only a small sample of the possible forces that might be generated are listed in the illustration I've provided.

Re: Increased homeowner water conservation in DesertSun

HELPING	HINDERING
Recognition of need	"My needs are unique"
Feeling of civic responsibility	Forgetfulness
	Enjoyment of greenery
Attraction for cacti	Tradition of using water
City ordinance	

Figure 5. Force field analysis example, step 3.

4. Once all items are listed, weigh each on a 0–10 scale, low to high impact, in terms of the amount each force exerts on the center line or present equilibrium. Some items may need to be clustered by theme before the weighing. Observe Figure 6.

   You could go through the forces again, this time rating them on a high, moderate, or low possibility of being effectively changed. The second rating would sharpen your sense of where you could overcome resistance and what you might want to let alone for the time being.

   Weighing calls for judgment. At times our ability to evaluate is good, at other times poor. Some of us are better at differentiating technical options, others at human factors. Those whose psychological type is "judging" prefer

Re: Increased homeowner water conservation in DesertSun

	HELPING	HINDERING	
8	Recognition of need	"My needs are unique"	5
4	Feeling of civic responsibility	Forgetfulness	3
		Enjoyment of greenery	8
2	Attraction for Cacti	Tradition of using water	9
5	City ordinance		

Figure 6. Force field analysis example, step 4.

arranging data and making decisions quickly, sometimes before all facts are on the table.

Others who are "perceiving" tend to play with data, holding off decisionmaking, sometimes until after a decision is due. Even the most logical person can be blinded by a pet peeve or idea. While experience may be a good teacher, we may be lousy learners.

Wise judgment is what is needed. It is important to trust our intuition and our experience, but it's also critical to heed rational thought. It's vital to know that blind spots can hamper our judgments. Thus armed, we can invite the views of others to balance our own. Making wise conclusions is as necessary when we are weighing our analysis as it is when we are choosing solutions from a host of ideas.

5. Review the forces and their ratings. Then come up with bright ideas to reduce, add, modify, divert, or delete forces. Even though force field analysis is an analytical method, this is where you step over into the search phase. Proposing ideas and arational thinking can, of course, enter in at any one of the phases.

Force field analysis is a handy tool for evaluating the positive and negative pressure applied by a variety of factors to keep a situation in its current condition. To make changes, the forces have to be identified, analyzed, and used to trigger ideas to alter the situation.

Cause/Consequence Analysis

Because interaction exists between causes and consequences, it is sometimes difficult to separate them or to determine their sequence. Graphically displaying causes and consequences can be especially important when dealing with physical processes and properties or when situational scanning is confusing or complex. Begin with either causes or consequences. One method is:

1. Start with a key consequence (effect) or cause at the center of a bubble. For example, you can create a map of possible causes by using spines rising from the bubble as causal themes, e.g., methods, materials, resources. (If you put a critical cause in the center, you would be searching for consequence themes and subthemes.) See the format in Figure 7.

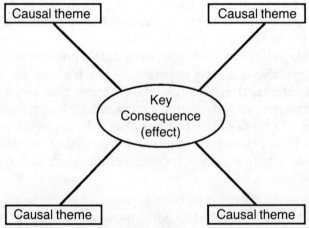

Figure 7. Cause/consequence analysis example, step 1.

Use additional spine branches to detail more specific causes and subcauses off each theme spine. Once the basic structure (bubble and causal theme spines) is set, be spontaneous. You don't have to carefully gather all the causes pertaining to a single causal theme before you move to the next one. Capture ideas as they come to you, jumping from theme to theme if that is what is happening in your mind. Let the structure build naturally, applying both intuition and logic to create the relationship of causes.

For the first example, we can go back to the case of the parched land. We could survey the causes of insufficient water in DesertSun by using the cause/consequence analysis. Scan Figure 8 to see how the study might look.

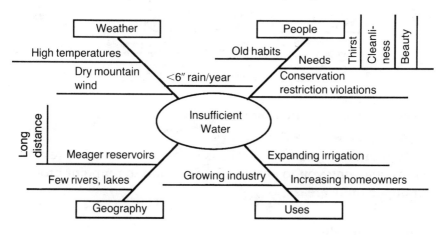

Figure 8. Cause/consequence analysis, example 1.

In a second example, customers—mostly builders and architects—of a company making wood-framed windows have returned an excessive number of windows from building sites around the nation. A typical investigation of such causes would center around at least four themes—people, machinery, methods, and resources or raw material. Figure 9 displays some causal possibilities built around the themes.

2. Determine the impact of each spine and branch. Is it of high, moderate, or low impact? Technically oriented, following laws of physical properties/processes? People-oriented, following laws of human behavior?

3. Look for gaps, departures from standards, and expectations. Determine areas that need additional investigation. Pinpoint areas that could be turned into targets.

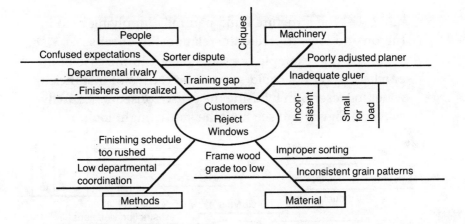

Figure 9. Cause/consequence analysis, example 2.

The cause/consequence analysis displays cause-effect relationships, allowing you to evaluate impacts and to pinpoint problem or opportunity areas.

Pareto Principle

Pareto discovered the 20%–80% principle, e.g., 20% of the effort produces 80% of the results, while 80% of the effort brings 20% of the results. You can have fun applying the principle to a variety of things around you. At work, 20% of the products in vending machines produce 80% of the profit. Twenty percent of the employees produce 80% of the productivity. Certainly there are limits and exceptions to this guideline, but it is a handy rule of thumb. One way of applying the Pareto principle is to use it in the following analysis process.

1. Create a list of items around a category or theme, e.g., a product line or work activities.

   Imagine you are an overworked project manager developing an experimental prototype for a new supercomputer. Your harried schedule erodes your quality of living. You rarely get out to play golf. It's been many months since you and your wife have been able to slip out of town together. You snap at co-workers, your kids, even yourself. Somehow you have to regain control of your life.

You start to manage your chaotic schedule by developing a list of work activities, as illustrated in Figure 10.

ACTIVITIES

Referee opposing team factions

Make presentations on project status

Manage critical path

Join lower level brainstorming sessions

Lead visitor tours

Sign all financial requests

Foster team building

Review technical reports

Attend demonstrations

Visit teams in their work areas

Track project's financial status

Fill out library requisitions

Figure 10. Pareto principle example, step 1.

2. Make five columns to the right of the list you develop. Designate the first column the "critical few" (where 20% of the effort will give you 80% of the work payoff) and a second titled "trivial many" (where 80% of your effort only supplies about 20% of the results).

Add a "must do" and a "want to do" column, to assess whether the pursuit is a requirement of the job or just personally satisfying.

A fifth column is for "hours per month" spent on that activity, the cost in terms of time given to it.

3. Make a check in the appropriate column(s) for each activity. A task should have only one check for the "critical

few" or "trivial many" columns and one mark for the "must" or "want" columns. If you are wanting to make more than two marks for each activity, look deeper at what is happening with that task. You should divide that undertaking so you can have a clearer evaluation of its impact, desirability, and expenditure of time.

After drawing the four columns, check the appropriate boxes for each activity. Add the hours per month given to the task. Study Figure 11 to see how the completed analysis might look.

You were so caught up in the forest that you hadn't stood back to survey the pressure points in your work schedule. You now know the activities that give you the most mileage and most need to be done by you.

4. Identify ways to reduce the "trivial many" and strengthen the "critical few."

Decide what activities most support your leadership role. Find approaches to drop or delegate the other monkeys from your back. Six of your activities were registered in the "trivial many" column. Some were musts, others wants. You could now go to the search phase to generate bright ideas to get your schedule manageable.

Look, for example, at how you gave an activity like making presentations a mixture of responses. That activity could be divided for clearer review. Proposing design changes to top management requires your personal, entrepreneurial touch. Other talks may not give you much bang for your time. Presentations seem to be required and you enjoy presenting your project to others. But that's also an area where you could allow others on your staff to develop their speaking talents and wean some departments away from depending so much on you.

Tough decisions may have to be made as you explore your activities, but give yourself room to think of the possibilities.

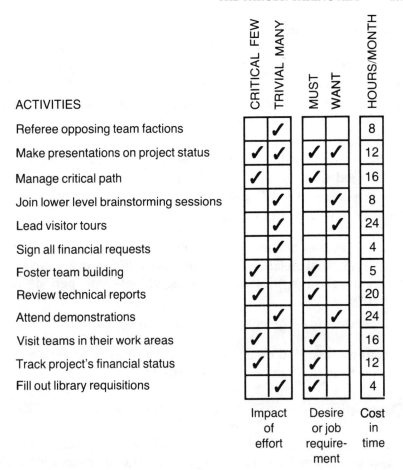

Figure 11. Example of completed Pareto principle analysis.

By using the Pareto 20%–80% principle, you can see where the "critical few" activities are giving you the most impact for the effort you put in. The "trivial few" tasks could become another staff member's growth opportunity, if you are willing to delegate. By observing the must—want category, driven by a job requirement or your personal desire, you can weigh how much you want to hang on to an activity. The hours per month column helps you visualize the time expenditure of each task.

Through this analysis, you can evaluate such things as activities, processes, products, outputs, inputs, and resources.

Historical Timeline

A sense of history is often instructive. Today's pressure can blur the accuracy of our memory for happenings of the past. It's easy to forget the sequence and impacts of events. As we drive into the future, however, we sometimes need to look in the rear-view mirror to see where we've been.

In consulting with both technical and business clients, I have discovered that developing a timeline is a very handy tool for giving perspective. We can extract personal, technical, and organization lessons from such an exercise.

1. List events, actions, or decisions in their sequence of time. Study their relationships. Or break a life span—of an organization, a project, or an individual—into periods or themes.
2. Review for problem or opportunity insights.

Creating a historical timeline can give you lessons of the past and perspectives that can be applied to the present.

Observation

Survey situations using a visual or hands-on approach, or both. Many inventions have been discovered by good observation techniques.

Be aware, however, that several tricks of the mind can provide sensory blinders as you observe. The ability to stuff data into pigeonholes and stereotypes can prevent your seeing with fresh eyes. Remember that another well-documented mental trap, the capacity to look for and see that which you expect, can imprison you in your expectations, in your "self-fulfilling prophecies."

Fleming was able to observe what others could not see. Others saw dead microbes around certain types of mold in a culture dish. But they couldn't intuit that it was from some substance emitted by the mold. He identified the specific substance and discovered penicillin.

Keep all your senses open as you observe. Follow these steps:

1. Get out to the assembly line, field, computer room, construction site, laboratory.
2. Observe the work flow, the interaction of people and the technical processes and systems. You might develop a chart to mark the number of errors, interactions, or compliance with expectations. Where appropriate, tabulate. When it's time, draw conclusions.
3. Observe nonjudgmentally.
4. Challenge your assumptions about what you see, hear, and observe, realizing that you may be blind to the unexpected.

By being alert, curious, nonjudgmental, out where the action is, and aware of relationships, keen observation will give you valuable data for understanding the situation and the target.

Value Analysis

Study the value you get from products or processes. Then explore the costs—dollars, effort, or resources—incurred to get those benefits. This analysis is the basis for an evaluation of cost-effectiveness.

I want to emphasize the type of value analysis that dissects functions to determine their role and benefit in a system. Are you concerned about:

- The effectiveness of bus routes?
- Where you could get the most mileage from robots on an assembly line?
- A bottleneck in updating engineering drawings?
- Choosing among a welter of medical procedures?
- Sorting out the interrelated and sometimes conflicting activities of a research team?

Using value analysis, a small team could give you a wide range of viewpoints about the systems or processes you want to evaluate. You'll want to tap people who can tell you something about the functions of the system—users, developers, inputters, workers, supporters, or sponsors. Your analysis is likely to include observations of actual operations, work flow diagrams, and statistical methodology.

Proper value analysis could entail days of effort. Time may be needed to collect and analyze specific data. You might, therefore, want to use an industrial engineer or other knowledgeable person to facilitate the collection, tabulation, and evaluation process.

1. Study the functions and benefits of systems or processes. Express a function as an action verb, such as *grease* machine, *collect* data, *unload* boxcars.

   Start with the function that begins or ends a process, moving forward or backward one function at a time. Make sure you link functions that form a sequence. Keep one-time-only functions (activities not a part of the main flow) separate. Determine the value that is added for each function. Assess accumulated values, too.

2. Explore the costs—time, labor, raw material—that are accrued to produce and maintain the value. Compare these costs with estimates of what the expenditure of resources should be.

3. Identify areas fruitful for generating other options. Look for the accumulation of small costs. Or target key cost areas that, if eliminated, would highly reduce costs or raise productivity, or both.

In value analysis, you define, arrange the sequence, and assign a value to functions involved in producing a product or completing a process. You spotlight functions with a high expenditure of resources. Then you generate good ideas to handle those functions in a more productive, less costly way.

▫ ▫ ▫

Some analysis methods we've just explored can be used interchangeably or can supplement each other. Often these techniques can be used in a "quick and dirty" fashion to rapidly get data on the table so you can look more intensively at specific areas. Or you can use these approaches in great detail to carefully and methodically probe a current state.

This is by no means the end of the analysis story. Numerous other books elaborate techniques to investigate the current and desired states. If you feel the need for additional detective work, turn to sources that provide a more detailed process.

One of those sources is Charles Kepner and Benjamin Tregoe's *The New Rational Manager*, in which you'll find several meaty analysis techniques. For studying the value of selected targets or goals, turn to a book like Ralph Keeney and Howard Raiffa's *Decisions with Multiple Objectives: Preferences and Value Tradeoffs*. If you want to discover a fascinating approach to analyzing technical systems, as well as gain excellent insights into applying creative thought, try a method that appears to work well in the Soviet Union. A Soviet scientist, G. S. Altshuller, believes he has unlocked the door to tying together imagination and technical problems. Reading his *Creativity As an Exact Science: The Theory of the Solution of Inventive Problems*, even though the translation or the writing style is rough in spots, could give you challenging new insights.

Situations need to be explored sufficiently before targets can be developed and proposals generated. The foregoing techniques help you gather, display, and assess data to illuminate current conditions. Analytical skills, therefore, are important in the creative cycle and should be honed.

ARATIONAL MODE POSTSCRIPT

On first blush the target phase may seem like a rational person's dream. You are asked to deal with facts and forces, with definitions and requirements, with winnowing data, with processes of the rational mind.

Don't put a straitjacket on your thinking, however. There are many places during the targeting process where you can exercise your arational mode.

- *Analysis needs the creative, too.* Situation analysis and force field analysis, for instance, require an unharnessed, imaginative flow of ideas. That first blast of opinions, perceptions, feelings, and factors you want to gather can be guided by the intuitive, as well as by logic. Using nonverbal symbols, word pictures, and metaphoric thoughts can add nuances, fill in gaps, provide a sense of the scope and intensity of issues, and give more of the big-picture focus.

- *Honing targets requires vision.* Vision bereft of imagination is often hollow. One of the creative thinking strategies, imaging the future (Chapter 7), is excellent for applying innovative thinking to the shape of targets.
- *Criteria may be produced in the form of wishes, hopes, and dreams to add creative spice, to both supplement detailed requirements and expand your thoughts.*
- *You need not stick rigidly to sequential development.* In gathering information about the situation and the target, you need not stick rigidly to the sequential development of the situation before you dare write down or draw ideas you have about the target. Capture ideas about potential target statements that float across your mind's eye while eliciting situational information. Although it is often helpful to finish analyzing the current condition before forming target statements, allow yourself to do what the natural mind does—to think in patterns and leaps rather than in a highly step-by-step process.
- *Feel free to use images, sketches, symbols, graphics.* These nonverbal expressions dramatize, underscore, or show relationships or depict the unexplainable. Restricting yourself to just words is like the old oriental custom of binding a woman's feet. Using only the verbal crimps your mental dexterity and movement, just as bound feet can cramp a woman's physical expression.

COMMUNICATION TIPS POSTSCRIPT

A technique is a technique is a technique. It's a guide, not an infallible procedure. There's no magic wand that goes with it to make it work. It's simply a process for getting a hoped-for result. Great chefs in the world's fanciest restaurants, as well as good cooks around the globe, aren't slaves to cookbooks. They deviate, try new concoctions and blends. Likewise, with any algorithm or heuristic, a human touch is often needed to make the creative thinking process come alive. Each one of the creative cycle phases has techniques. They require the logic of the rational and the imagination of the arational.

The human touch comes alive when you facilitate the flow of a meeting using the creative cycle and assist the interaction among the members. Facilitation requires using effective processes for easily moving from step to step. These approaches free people to communicate and produce good ideas while keeping the group moving in a productive direction. At the heart of this expediting is promoting an alert, sensitive, trusting exchange of information that yields imagination and shared understanding.

As soon as more than one person becomes involved, the level of communication will make or break these techniques. Here are some ways to aid effective communication and facilitation during the targeting process:

- *Get to the people who have the data.* Sometimes information is collected in a vacuum, without the people who can really provide input. Individually interview or have a small group interview of people closest to the action—people on the production line, marketeers, craftspeople, customers, internal and external clients, sponsors, kibitzers, secretaries, individual contributors, different levels of management, insiders, outsiders, servers, users, those concerned with detail, as well as those who can give the big picture.

  Either a thunderous roar or a whimper can be heard from one of the innovations of the workplace—quality circles, quality of worklife groups, or quality improvement teams. These innovations can be a problem solving force in companies or a dud. These groups can take big leaps for organizations when they get people who are close to the problems and can give a balanced perspective. The wrong mix of people will cause flat results. Of course, members of these

114   CREATIVE THINKING AND PROBLEM SOLVING

teams need to be trained in identifying issues and finding good solutions. And they have to have the communication skills described below to balance their problem solving.

A broad spectrum of people can help you create a storm of situational factors and targets that can set the right foundation for determining the current and the desired state.

Be cautious about who you involve in this and the ensuing phases, however. You don't want stick-in-the-muds, people so enmeshed in the situation that they come up with a thousand reasons for why something doesn't or won't work. They could be lead weights on a team, causing data collection and creative thought to sink.

- *Create a mood for synergy.* Let people know you care what they think. If you really believe in synergy when people put their heads together, you'll show it in your face, body, and behavior. When people suggest an idea for how 2 + 2 can equal 5, 9, or 20, applaud, rather than cock your head at them with disbelief and disdain. Go to Chapter 10 for suggestions on how to strengthen this mood.

- *Effectively use questions.* Open-ended questions can be a key to getting good information. They are the kind that produce a variety of responses, rather than a yes, no, or I don't know. Have you stopped beating your pet barracuda? calls for a yea or nay. Is your manager still as messed up as ever? leads your opinion giver into a corner. However, What are the key elements to be considered? or How is the professional staff affected by the new appraisal system? or In what ways will the proposed waterway affect the surrounding area's ecosystem? allows for a range of answers.

  At other times, you must ask pointed, specific questions to carefully gather the facts in a situation. Specially shaped questions can be very appropriate, such as when you are in pursuit of the cause of a technical malfunction or error. Provided you don't put words in someone else's mouth or harass the questionee, directed questions are especially useful in a method such as situation analysis where you must elicit data about how the critical factors in a situation deviate from what is expected.

Asking the right type of question is often the key to unlocking a confusing, complex, knotty situation.

- *Honor the differences among people.* Let people with other perspectives and from other disciplines know you honestly value the unique lens through which each person looks. While this attitude is absolutely critical at the second phase of the creative cycle, it is vital here, too. Collecting data means honoring, not browbeating, sources. When people believe their viewpoints are discounted, they may tell you only part of the facts or truth.

    Thus, when you have egg all over your face, someone slips up to you and says, "Oh, we forgot to tell you. All that data you used in your presentation to the bigwigs upstairs was from uncalibrated instruments!"

- *Make issues and data visible to yourself and others by writing them on flipchart paper or in a journal.* Even if you have a photographic memory, write thoughts down. Viewing ideas, either in verbal or image form, often stimulates additional ideas. When working with others, listing options will underscore your concern for their thinking. Posting the lists on the wall will give you and the group an accessible, collective memory.

- *Blend task with process.* That is, blend activities that highlight *what* has to be done with activities for *how* something will be accomplished. Summarize. Note the time remaining. Clarify the content. These are helpful tasks. Suggest the development of a flow chart. Choose an analysis method. Ask your team what would be more effective ways for working together. These are processes that help accomplish the task.

Although these communication tips are presented during the description of the targeting phase, each one should be used in the other phases, too. Effective communication skills provide a useful foundation for the whole creative thinking process. They are especially helpful in any group creative effort.

We have reviewed the various steps in the targeting process, added a variety of analysis methods, noted ways to enrich the rational with arational approaches, and highlighted a few communication tips to more effectively pull off this phase. Now we can explore how to search for ideas.

6 Search Keys: Igniting the Imagination

> *Here's the World War I flying ace walking out onto the aerodrome at Rembercourt.... I have been assigned to a night-fighter unit . . As I climb into the sky, the huge LeRhone rotary engine in my Sopwith Camel throbs its song of destiny.... My mission is to fly south from Verdun to St. Mihiel, and then southwest to Bar-le-Duc, hoping to trap a German Gotha bomber in the night.... There's only one thing wrong.... I'm afraid of the dark!*
> —As Charles Schulz had Snoopy say in a classic PEANUTS strip.

> *Let go.... Let it happen.* —TIMOTHY GALLWEY

Searching for ideas isn't new. Over the years we have had to come up with many responses to situations that have taxed our ingenuity. How to sneak into the house way past dinnertime with a good excuse for our folks or our spouse? How to give our teacher a sound answer to a question asked while we were staring out the window with a mischievous daydream? What kind of a prank to play to get even with the joker in an office down the hall?

Ken Blanchard, author of *The One Minute Manager* and other "one minute" books, told Joel Goodman, editor of *Laughing Matters*, about his search for an idea in a touchy situation in his youth.

Ken recalled when his sixth grade team was playing for the championship. A member of the other team, who seemed like he was well over 6 feet tall, was dubbed Meatball by the other kids.

After Ken played well and his team won, he went into the locker room to change. Ken passed the tall lad from the opposing team who was sitting on a bench, patted him on the back, and declared, "Good game, Meatball." Meatball grabbed him, threw him against the lockers, and said, "Only my friends call me Meatball." After a very quick mental search, Ken said, "Well, then, why don't we be friends?" The tall youth laughed heartily, saying, "You're all right, man." Sometimes we need good ideas in milliseconds.[1]

Life is built on figuring things out. We may not need to produce such a quick, humorous retort as Ken Blanchard used to get out of a delicate situation. Yet we often need good ideas to succeed in daily life. To taste success, most engineers and scientists need a stream of bright and novel ideas. Some of the ideas require spontaneity, others more thoughtful attention.

In this chapter we will explore the overall strategies you can use, as an individual or in a team, to ignite the imagination. These are approaches to shake up, tear, unlock, splinter, or evaporate those mind-sets that keep us blind to new possibilities. We will also examine some of the critical elements that carry out the strategies and lead us to a creative state of mind, generating rich, fresh, breakthrough ideas.

MIND-SET BREAKING STRATEGIES

Searching for ideas is a matter of altering the usual way you look at a problem. Breaking mind-sets is a key task of creative thinking strategies. The more imaginative you want your end product to be, the more you have to stretch and smash stereotyped ways of thinking.

If you feel strange when tackling some of the tasks and processes that lie ahead in the search phase, that is probably a sign you are being pushed in new directions. That is the point. You may need to jar or tease or ease your thinking into new territory. You may have to try the less familiar to break a mind habit and open up innovative options.

Mind-bending strategies are deceptively simple, yet require continued practice and refinement. The strategies can be considered techniques, yet they require personal involvement and a proper mental attitude. They include elements that can be puzzling to concrete or logical minds, but which open the door to improved rational, as well as arational, thinking.

To prepare yourself for understanding the three mind-set breaking strategies, imagine an aikido martial arts expert practicing and meditating. For the aikido expert to release energy, several things need to occur at the same time. The master centers on inner strength, relaxes mentally and physically, taps wisdom of the mind, senses his whole body, and moves spontaneously but purposefully. He is a bundle of flowing energy.

To release creative energy, three things—storm, calm, and access—need to be combined. They may seem to be contradictory, but should occur in concert for the most effective flow of imagination.

Storm

Storm means *provoking a wild, unpredictable storm of ideas*. Most of us know what it is like to get caught driving through an unusual summer storm—when heavy rain buffets us, swirls of wind sway us, leaping lightning bolts and thunderous clouds awe us. There is great energy being unleashed in a storm.

To produce good ideas, we have to create such a storm. We need to release the energy of our thinking, our attention, our enthusiasm, our emotions, our imagination. Bold, adventurous thoughts that blow us away from staid thinking have to be permitted.

Many people approach idea-generation sessions with the caution of a gentle drizzle while the sun is partially shining. They don't want to get caught in any event where they feel they may get out of control, so they allow only a trickle of ideas to flow. Really creative thinking requires us to get into the thick of an idea storm.

Calm

Calm means *inducing relaxation. Reducing stress. Allowing ideas to percolate in the subconscious.* This is paradox—calm and storm at the same time. Calm is needed to gain the benefits of storming. In this strategy, soft focus is held up as a goal, as a way to allow stimulation and relaxation at the same time. A peaceful, self-possessed attitude allows storms of ideas to swirl in one's head or in a meeting without worry about the potential outcome of options.

A mind that is relaxed and approving sees possibilities in ideas that are fragmentary, a bit crazy, or off-the-wall. This state of mind is curious, positive, and confident. It is a mental attitude and state that builds up or redoes ideas to make them winners, rather than readily tossing them out if they seem strange or a bit off the mark.

Access

Access means *tapping intuitive and preconscious thought.* This strategy gets us in touch with potent creative problem solving that takes place out of our normal awareness. Intuitive, preconscious processes often seem to be hidden in the forest of our minds—darting into a clearing to tantalize us with an insight, then disappearing into the forest before reappearing again in some unexpected way.

Link calm and access. Use a quieting of the mind to allow the intuitive to be heard. Deliberately use periods of meditation, dream sleep, and that floating time between being awake and being asleep as an opportunity for imaginative thought generation.

What you don't know about your mind can hurt you or at least hurt your ability to allow good ideas to bubble to the surface. Skepticism of that which can't be observed with the senses or logic keeps a lid on the wellspring of innovative thought.

The three strategies work in tandem. You could put more emphasis on any one of the three to meet a particular preference. But rich, breakthrough thinking requires an abundant supply of each one. Figure 1 suggests not only that the three strategies are interrelated, but that the more they are combined, the richer the potential for getting top-notch ideas. We'll dissect these broad strategies here. Then, in the next chapter, we'll fold them into specific techniques that can help you hunt for ideas.

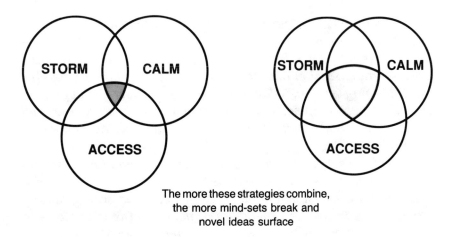

The more these strategies combine, the more mind-sets break and novel ideas surface

Figure 1. Scant vs. strong combinations of the search strategies.

SEARCH KEYS

To create a storm, bring calm, or access intuitive, preconscious processes, you need more than wishful thinking. You need to discover and use the crucial ingredients that make the strategies real, that ignite the imagination. Allow yourself to fully explore the search keys presented here.

Some of the keys underscore the storm, others help bring the calm, while others aid the access. These imagination sparklers can be used solo or in a team.

Spark a Storm

This technique is a cornerstone of generating good ideas. Through his book, *Applied Imagination: Principles and Procedures of Creative Problem-Solving*, Alex Osborn popularized the verbal brainstorming process during the 1950s. Although Osborn geared his technique to groups, you can also use it solo to produce a series of solution alternatives. Storm in a team. Storm alone. The payoff can be gratifying to groups or individuals.

Robert Bailey, the electrical engineer and creativity guru, was teaching his university students how commutators worked on a direct current motor. Often he started his classes with a 5-minute brainstorming session with the whole class. This time he had his own storming session, wildly suggesting a new, radical approach. Bailey proposed replacing the commutator bar with a solar cell and having a solar shaft motor. Thus his solar electric motor was born.[2]

Although still in the embryo stage and awaiting further refinements, perhaps even additional discoveries, the solar motor concept teases the imagination with possibilities for an energy hungry world. A sketch of the experimental model that was tested is in Figure 2. This one instructor's freely poured out ideas in front of a class is just one example of using individual storming.

To spark a storm, use the question prompt you developed in the target phase. If crafted well, the prompt can be the match to get the blaze of ideas going, for it contains the direction to focus your search.

Many people have a narrow view of storming. Storms may go far beyond listing words that an individual or group generates. Drawings, symbols, metaphors, or mental imagery can be used to provoke, as well as express, the flow of ideas. The more sensory modes you combine during the storming, the more creative power you will be able to generate.

If you want the most innovative thoughts possible, first get the commonplace ideas into the open. Start an idea-generation session with a quick verbal storming (4 to 5 minutes), thus draining the tamer or more practical ideas. In some circumstances, this

Figure 2. Experimental model of first solar electric motor. From R. L. Bailey, *Disciplined Creativity for Engineers* (Ann Arbor, MI: Ann Arbor Science Publishers, Inc., 1978), p. 246. Reprinted by permission of Butterworth Publishers, Stoneham, MA.

burst of the more common ideas may give you sufficient options with which to work. If you want to push on to get more imaginative ideas, turn to longer nonverbal storms or other techniques which give you a greater chance of getting more original options. It is vital to allow time for such ideas to occur.

In the next chapter, a number of storming strategy formats are presented. You may also build your own by incorporating some of these key search ingredients into a tailor-made sequence that can have creative payoff.

Follow Storming Rules

Use of well-researched rules permits an increased, unhampered flow of ideas. Misuse of these rules can cause the storming to go from a potential monsoon of good thoughts to only a light drizzle that doesn't quench the need. The rules:

- **Use no judgments and no discussion while storming.** Violate this at the peril of killing even tame ideas, let alone those options that can push the boundaries of thought. This is a difficult rule to uphold for most individuals. Many technical people, who have been taught to carefully scrutinize, analyze, and weigh data, have a hard time putting those skills on hold for a while. This rule is vital to openness of reasoning, whether you are producing thoughts alone or when in a storming session with others.
- **Welcome wild ideas.** Many storming sessions suffer from not enough wild ideas. Zany, bizarre, or off-the-wall comments don't mean that you are going off your rocker or out of control. Ideas that bring belly laughs or cries of "craaazzy!" don't just stimulate bolder options or break thought boundaries. They often contain seeds of brilliance. With nurturing, many wild ideas become winners.
- **Quantity is important.** Get all the ideas out. Remember scripture. Remember the "begets." One idea begets another and another and another and . . . Use the bunny rabbit habit. Let ideas stimulate the reproduction of more thoughts and more and more and . . . Some techniques, like *analogy storm* or *picture tour* described in the next chapter, however, may cultivate and expand only a few thoughts until they are rich and novel. The idea is essentially the same: let bright ideas—one or many—continue to unfold until the spice and sparkle of imagination has run its course.
- **Piggyback on the ideas of others.** Enjoy a ride on someone else's thoughts. To build on someone else's good thoughts is to use that thought as a stimulus—and to pay that person a compliment.
- **Honor and push beyond silences.** Ride the silences, too. Incubating is going on during moments of quiet. Are you one of those with a tendency to be decisive, to quickly tie a ribbon around a thought, to go for the bottom-line juggler? If so, allow yourself or the group to hesitate. Research has shown that the best ideas come after the second and third silences.

Tinker with the Angles

There are times that just glancing at something from a fresh vantage point can bring an *aha*. Look at a problem, product, process, vehicle, or relationship from different directions. Shrink, stretch, reverse, raise, remove, or lower it. See what new insights can be derived from that new perspective. Gain new viewpoints, for instance, by going backward rather than forward when looking at something.

Much of the airline industry has tried to advance by plunging ahead—moving steadily to develop jet-propelled craft. This advancing has caused them to tinker with the perspectives at which they look for ideas, looking over their shoulders to the past to be able to move into the future. Because of fuel gobbling jet engines, some airplane companies have reversed their viewpoint, going back to the propeller-driven, more fuel-efficient plane. New propeller designs are enhancing the concept of going backward to go forward.

Another practical way of looking at something from a reversed or new angle is to create a storm of the things you *don't* want something to be. This different perspective can help you then imagine ways to prevent those deficits from coming to pass.

Stimulate and Relax the Senses

Remember the case for using the primary sensory media—vision, audition, touch—in breakthrough creativity efforts? As you rummage through your mind in search of ideas, make free associations while you stimulate yet relax the senses. You are after a soft focus approach.

Use paintings, sketches, photographs, or words and symbols written on newsprint during a storming session. Use the imagery of day visualization, described in Chapter 2, and night dreaming, discussed later in this chapter. Explore problems and solutions using your powerful image-making mechanism. To practice freeing up the visual, nonverbal mode, you might want to obtain Betty Edwards' Drawing on the Right Side of the Brain.[3] *She has a number of exercises you could try.*

Listen to meditative or baroque music and to such environmental sounds as waves lapping against the shore or wind rustling through treetops. Try nonblaring but emotionally stimulating music that can cause the whole of you to be alert and receptive but relaxed. Walk on sand with your bare feet. Sculpt soapstone. Mimic the movements of a process, animal, or object with your body. Take a shower to awaken tactile feelings. Make something in your woodworking shop, feeling the wood as you go and letting your imagination roam free. You could also try special smells and tastes that could enkindle a creative environment, as well as excite the imagination.

As we continue this exploration of the keys to the search, you will note that the senses are also a keen part of a number of the other crucial ingredients.

Use Analogies

Word pictures or analogies are as common as a morning cup of coffee or brushing our teeth. Many people sprinkle their conversations with them: "Can't get the old motor running yet." "He listens as if he had a bale of cotton in his ears." "Being around her is like walking through a mine field." "I'm as low as a snake's belly."

By creating such analogies when you are problem solving, you can get clues to possible solutions. Inventors, scientists, and theorists have, throughout the centuries, fired their imagination with analogies. They received *aha*'s by comparing a concern rolling around in their minds with an object, life form, system, process, or element.

William Harvey was observing the gory, exposed heart of a live fish when an analogy created his aha. He began to imagine the pulsating organ as a pump, with blood moving through the body of a fish like fluid flowing through a mechanical pump. At first ridiculed, his discoveries of the circulation of blood greatly boosted the understanding of the human body in the seventeenth century.[4]

Through analogies, two men found the solutions to different problems on adequate seating for the physical body—one for Earth, the other for outer space. Engineers struggled with how to design well-contoured seats for a new sports car being developed by Ford-Europe. They had tried expensive mechanical models that still didn't do the job effectively. Their new head of programs and marketing, Bill Camplisson, was puzzling over the dilemma one night, when he thought of being on the beach as a child. His beach ball had gotten crumpled when someone stepped on it. Bill's father came over to comfort him and, taking the ball into his hands, popped the ball back into shape. That reshaped ball from prior years was the eureka, the analogy that was able to guide the design for comfortable bucket seats.[5]

Maxime Faget wrestled with another seating problem. He was concerned with providing astronauts with a seat that would allow them to perform their many tasks—such as handling the control panel and thinking clearly, while also withstanding g forces against their bodies. His inspiration came through an analogy. Eggs, with their delicate outer shell and flexible mass inside, could be cradled to withstand much pressure and also protect their inner contents. From this analogy, Maxime was able to design an effective, protective astronaut's couch.[6]

You—like Harvey, Camplisson, and Faget—can take a characteristic from an analogy and shape a solution with it. Approximation and adaptation is the name of the game. If you find at least one similarity between the things you are trying to compare—like the delicate contents of an egg to an astronaut's body, there may be agreement on other aspects—like cradling the contents of the egg and body. Tantalize your imagination by searching for metaphoric connections to your problem or hoped-for outcome.

Change Your Form

Let your imagination create a mental metamorphosis. Take on imaginary properties. Become objects or processes in the situations you are studying. Feel how they work or are put together or interact.

When concerned with insects attacking a tree, for example, you might begin, "I'm the branch of a tree. I feel very naked and *vulnerable, for small creatures are crawling over me, pulling chunks from my body. I wish I could brush them off or prevent their digging into my bark with their pinchers."*

Allow someone who is trained in guided imagery to help project you into a changed form, into a structure similar to the one in your problem. Visualize the altered form. Make vivid any kinesthetic sensations. Listen to sounds and to what even inanimate objects and processes might be saying. Smell odors. Note tastes.

Force Fit

Heavily rational people may think those of us who teach creative thinking have a screw loose when this approach is suggested. Yet, the forced fit is active in a number of the more powerful creative storming techniques that produce novel solutions. For example, force fit is one of the active ingredients in using analogies.

Force fit is the mental process that calls for you to make connections between things that are seemingly unconnected. For example:

Take a benign thought like "reflections of the sun." Use that neutral phrase to intuit solutions for preventing prepackaged foods from becoming too moist in microwave ovens. Use the "anus on the back end of a farm animal" for creating a way to transfer food from a container to a mouth in outer space. Squeeze solutions from the "spray of a waterfall" to awaken workers who dangerously fall asleep on a conveyer belt during a lull in their shift.

All three of these problems, two from published literature and one from my workshops, found potential solutions in the "unconnected."

Take a Leap

Ask, "What's the most fantastic, bizarre idea I can think of? What's something no one else would dream up?" Take a leap to that which may seem far out. There's often a glimmer of genius in wild ideas that will push your thinking.

Wilson Greatbatch was a believer in taking a leap. He felt that you may curtail your thinking if every step you take has to be carefully laid out like a precise experimental design. In inventing the pacemaker, he took his "big jump." Wilson threw some wires together, using them to jolt a dog's heart to see if he could get it to beat. Refinements, such as encasing the device in material that wouldn't cause the human body to repel it or developing the best circuitry, could come later.[7]

Storms require leaps of thought. Timid thinking usually won't provide breakthroughs. The storming rule *Welcome wild ideas* is designed to trigger brazen broad jumps, rather than the mincing steps of mental hopscotch.

Extract

Sense the essence of a problem, situation, process, concept, or idea. Extract the key elements or concepts by storming, intuiting, or analyzing. Once the essence is visible, you can have further rounds of idea generation.

Structure

Once key elements have been extracted, you may want to lay them out for easy viewing. Organizing data—that generated through analysis or through a search technique—can help you see relationships between elements and options or show you gaps that need to be plugged with additional ideas. One good way of arranging data is to develop a matrix; another is to develop a morphological box.

Matrix

A *matrix* is simply a vehicle for breaking a problem into parts. On a matrix, you can compare such dimensions as subsystems and options, influencing factors and analogy worlds, subelements and mechanisms, or boundary conditions and bright ideas. It's a way of laying out a problem or opportunity so you can see gaps, connections, or places to look for new ideas.

We could apply matrices to such areas as designing power packs for outer space, identifying a new chemical in the brain, or developing high-speed computers. To give us a simple illustration, however, let's briefly explore the subject of . . . well, mice.

The mouse continues to attract attention at different places in the world, as an exalted research tool or as an unwanted pest. Ralph Waldo Emerson once remarked, "If a man makes a better mousetrap, the world will beat a path to his door."

Someone once told me of the mice plaguing his oceanographic laboratory. One day he opened his desk drawer and found two mice scampering around inside. He grabbed a cat and put it inside the drawer. There was feverish scampering around. When the dust settled and the drawer was reopened, the cat was standing with legs spread, the two front paws each on a mouse. The cat's eyes were bulging with a look of, "What on earth do I do next?"

At that location, cats were used to control the mouse population. Sometimes other gadgets are needed to get rid of the pests. Have some fun and imagine you were commissioned to produce the ultimate mouse trap. By doing an analysis on the functions of the mousetrap system, you might identify three basic elements or subsystems, such as *attract, capture,* and *eliminate.*

As a next step, develop a matrix with the subsystems in the left-hand column. Reserve the columns on the right for options. Then generate a number of options for each of the subsystems or

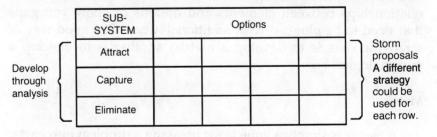

Figure 3. A matrix for arranging search data.

parameters. See Figure 3 for a suggested matrix. Turn to Chapter 7 for ways to develop options for the subsystems. You could then use a different strategy format for each row, depending on how rich and varied you wanted the proposals to be. Of course, you can generate as many columns and rows as you want.

Morphological Box

A matrix is two-dimensional. Astrophysicist Fritz Zwicky goes further. He suggests a three-dimensional structure, called a *morphological box*. He uses a disciplined approach to gain insight. By applying his technique, for instance, Zwicky has mapped extragalactic space, making numerous discoveries, including the clustering of galaxies and the luminous bridges connecting galaxies. Although his morphological analysis could have been listed in "Analysis Methods" in Chapter 5, I'm including it here as a structuring key for rummaging for new ideas.

To give you a flavor of his technique, I'll summarize his process, drawing from material in his *Discovery, Invention, Research: Through the Morphological Approach*. Here's how you could apply the morphological box to an exploration of the conversion of energy:

1. *Define energy as thoroughly and accurately as possible.*
2. *Pinpoint all the parameters of energy.* Identify the basic elements or types of energy, such as kinetic, chemical, elastic, heat, magnetic, atomic, gravitational, electrical, light, resting. Name the ways energy can be transformed from one form into another, such as chemical into kinetic or heat or another kind of chemical energy.
3. *Build a morphological box, with the types of energy on one axis and energy conversions on a second axis.*[8] You are now ready to add a third axis—devices that bring about the conversion of energy from one form to another. There can be a vast range of such devices—the taut bow that catapults an arrow, wind being harnessed by the sail of a boat, a wheel being turned by rushing water, the ignition of gunpowder and fusing of nuclear material, light turning photographic emulsions black, magnets attracting metal ob-

jects, fireflies blinking on and off. The list seems endless. That's why structuring helps to display the known and show where there are gaps of understanding or where devices could be expanded with creative thought.

Figure 4 shows a simplified version of a morphological box to portray the concept. All three dimensions have more categories, of course. And the dimension-depicting devices could be divided up in many ways. The figure includes an example of a few devices for harnessing natural kinetic energy and converting it into other kinetic energy.

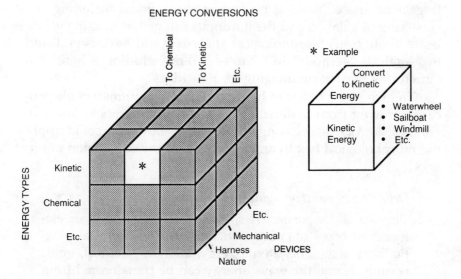

Figure 4. Morphological box showing relationships between energy types, energy conversion, and conversion devices.

4. *Analyze all the devices.* Comb them for holes, for places where energy conversion mechanisms are scanty, inadequate, or ripe for further development. If you find gaps, search for new possibilities and combinations. Intuition, as well as logic, is useful in this exploration, for you are looking for patterns, relationships, or the picture that can lead to insight.

If this morphological analysis seems like a large task, it is. That's Zwicky's desire. He's intent on surveying large chunks of knowledge to see if there are holes in our understanding and to discover new opportunities. You can take his full-blown approach, or you can borrow the concept to tackle a smaller but multidimensional arena. Or you can fall back on using just a matrix to launch your search for new possibilities.

Access Intuition and the Preconscious

Tap those marvelous "out of conscious awareness" mental processes that often linger out of sight or on the outskirts of the conscious mind.

Jot down thoughts right before you fall asleep or just as you awake. Awaken yourself after creative or interesting dreams to write down key thoughts from the dream. You say that is impossible, for you are too sound a sleeper? You say it's fruitless? For inspiration, read Delaney's Living Your Dreams *and Garfield's* Creative Dreaming. *You can also try these steps:*

- *Believe that your dreamland is a fantastic landscape on which to explore your imagination and to probe for solutions to important problems.*
- *Give yourself a dream assignment—telling yourself before you go to sleep to find solutions for some problem at work or in your personal life. Wake up at the end of a dream. Keep your eyes closed while you rehearse the key elements in the dream. Jot down those elements on a pad by your bed, and then write out the dream in the morning, as if it were just happening to you.*
- *Comb the dream for themes, symbols, and meanings that can help you find answers to on- and off-the-job situations.*
- *Practice, practice, practice.*

Buy a book with no lines and no words. This may look like an empty book, but your imagination can soon turn it into a masterpiece of creative thinking. Put ideas in this journal as they occur to you at odd times during the day. Each day intentionally relax with journal in hand and begin to write down random thoughts about topics or issues that seem important at the time. Free associate, letting random thoughts tumble about your mind.

Incubate

Take time to let ideas germinate, without trying to yank them into the open and pound them into shape.

Take a break from focusing hard on the target. Break for a couple of minutes or a couple of days. Take a walk. Hoe in the garden. Take a snooze. Go to a movie. Change tasks.

After the relaxation or change of pace, you will be able to refocus more sharply—and may find that some ideas have grown and are waiting to be cultivated.

The search phase attempts to stir the imagination, using strategies and keys to unlock creativity. Twelve critical elements for igniting inventive reasoning have been presented. They are summarized in Table 1 to give you a capsule description and the strategies they facilitate.

Table 1. A Summary of the Search Keys

Key	Description	Strategy Facilitated
Spark a storm	Using methods to elicit ideas	Storm, Access
Follow storming rules	Following norms for self-induced or group storming sessions	Storm, Calm
Tinker with angles	Looking at something from different vantage points	Storm
Stimulate and relax the senses	Arousing the senses while putting the mind and body into soft focus	Storm, Calm, Access
Use analogies	Looking for solution clues in analogies	Storm, Access
Change your form	Getting inside a process or object by taking on its properties, as if it could talk, think, behave	Storm, Access
Force fit	Connecting the unconnected for novel combinations	Storm, Access
Take a leap	Pushing the mind to accept big jumps in thinking	Storm, Access
Extract	Pulling the essence out of something	Storm, Access
Structure	Arranging data to gain insights into patterns or the need for further ideas	Storm, Access
Access intuition and the preconscious	Tapping processes usually out of awareness	Access
Incubate	Letting ideas germinate	Calm

THE BEGINNING

Now that you are armed with many keys for sparking the imagination, the fun begins. If you want, you can put together a technique or two of your own for stimulating creative thinking. Based on these strategies and key elements, you can concoct your own combinations and come up with unique methods for making innovation sizzle. Whatever actually works for you is right.

Or you can move to Chapter 7. There you'll find the steps of the search phase and some of the best strategy formats ready to be plugged into the creative cycle.

7 Search: Following Steps and Strategy Formats

> *What you see is what you get. Change your eyes.*
> —SAM KEEN

Hunters come in all sizes and shapes, with diverse methods and quarry.

A young lad on the prairie in the West takes shots at ground squirrels with a BB gun. His folks on safari get dressed in outlandish outfits, are directed by a professional guide, carry high-powered rifles, and stalk rhinos in the backlands of Africa. Clamoring shoppers rummage feverishly through mounds of marked down clothes at a large department store sale in a frenzied hunt for just the right bargain. Investigative reporters pour over piles of records in hopes of finding a clue to a potential government scandal. A diver with a net carefully swims her way under water along reefs of coral in pursuit of new species of ocean life.

During the search phase, the hunt is on. The quarry is ideas. This phase of the creative thinking process entails a wide variety of targets, searching in the right place, excitement of the chase, and the proper methods for ferreting out the rich ideas. In the previous chapter, you were armed with the key search tools.

We now stress proper methods for the hunt. In this chapter, those key elements of the search we just explored have been folded into practical formats to be used in stalking creative ideas. The approach you take in your search is where Sam Keen's quote can come alive. *The steps of this phase and the strategy formats help you to "change your eyes" during your pursuit of the novel and imaginative.*

We will examine the steps in the search, then explore techniques that can help us seek creative solutions. Before we close the chapter on prized search methods and processes, we will look at communication tips that are especially important for this phase. Figure 1 lays out this portion of the creative cycle.

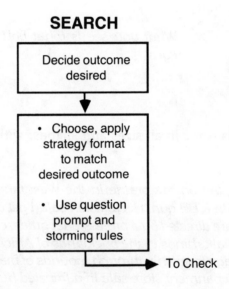

Figure 1. Search process.

You may like to hunt alone. Or you may prefer to ferret out your quarry with other people. The hunting tools I'm presenting in this chapter can be used solo or in a group. Having a guide or facilitator will help you get farther, however, especially when using analogies, pictures, images, and other nonverbal storms in a group setting.

As a youngster, I remember being out in the woods in an early attempt to hunt rabbits. A few others in our party were to come from another direction to beat the bushes, trying to flush out the prey. When a rabbit started down a path toward me, bullets sud-

denly started whizzing all around me. People not in our group invaded our circle, shooting indiscriminately, poorly. There was no one to guide the whole effort out in the wilds, just a lot of individuals haphazardly moving around.

Facilitators can help provide focus, steer the hunt, guide the process, and assure that participants aim at the target, not fellow team members.

STEPS IN THE HUNT

The steps in the hunt are only two, but each strategy format involves a series of things to do. Here are the two steps:

1. Decide Outcome Desired

> **Ask, "What am I after in this search? Am I primarily interested in an adaption, refinement, or better twist of something that already exists? Do I want to gain perspective or to scope an area? Is it important for me to find something that breaks new ground?" Reach a decision so you can gauge the effort, time, and degree of mental flexibility you will need to gain the outcome you desire.**

You may have decided the type of outcome you want before you started the whole adventure with the creative cycle. That's okay. Now that you have completed the target phase, reaffirm your earlier decision.

You may decide that you'll be happy with a solution that stays within limited boundaries. That can be okay. The crunch of a deadline is approaching and you don't have the resources or the time to stimulate, cultivate, shape, and test potentially novel ideas. Or, the people you have to work with are suspicious about or not ready for the methods that bring more original results. Yet you still need strategies to help you find good ideas that let you live within your constraints. You may not need a breakthrough, but you may need to get out of a rut of thinking or to receive a little boost of ideas to get you going.

If you choose to really stir the imagination and search for more novel, boundary-pushing proposals, you need to assure yourself you are ready to proceed. Processes for stimulating the more inventive require using arational thinking, playfully using such things as metaphors, music and movement, imagery of the mind's eye, and pictures. The strategy formats you need to use may take a bit more time and can be made more effective if you use a professional facilitator. You have to really be ready to handle the new perspectives, the different approaches, the proposals that push against comfort zones.

2. Choose, Apply Strategy Format to Match Desired Outcome

> Decide which strategy will produce the results you want. Start with the question prompt. Apply the storming rules. Follow the process for the specific storming strategy you have chosen.

Each strategy format, a packaged technique, has a certain amount of power to ferret out creative ideas. I use two categories of creative power, lower and higher. They are contrasted schematically in Figure 2, along with examples of techniques I'm suggesting.

Lower power techniques potentially produce moderate imaginative energy to create useful, better twists, and sometimes very unique thinking. These processes can be fascinating, produce many ideas in a short period of time, and have those who participated in the process feeling that they have contributed and gained something valuable. The majority of problem solving situations you face can be handled with these techniques. Several storming strategies fit into this category. We'll highlight three—traditional brainstorming, brainwriting, and mindmapping. And we'll look at the steps and an example of each.

Higher power strategies use more arational processes and techniques that result in novel approaches, producing stimulating creative energy.

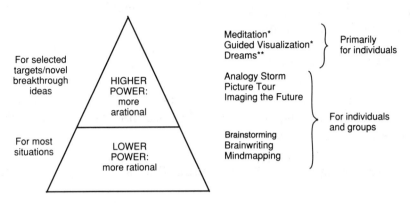

Figure 2. Higher and lower power strategy formats.

By releasing more creative thoughts, these methods have the potential for pushing you into newer areas, bolder combinations, and changing paradigms.

They require a good climate and a jumper-cable question prompt to get you going, the same as lower power or better twist methods. But higher power formats require more nurturing, more freedom of thought, more setting of the environment, more willingness to use arational media and modes, and, at times, more discipline in applying the process. In this chapter, we'll highlight three such formats: analogy storms, picture tour, and imaging the future.

In Figure 2, I show three other high creative energy techniques that can lead to novel ideas: guided visualization (Chapter 10), meditation (Chapter 10), and dreams (Chapter 6). Those methods can be potent formats for stirring the imagination and require a disciplined approach to milk their potential. When not consciously used, these processes carry on their creative problem solving play in the background, out of awareness. Additional good, but less structured, creative thinking formats can also be found in those chapters.

More formats of low and high imaginative power exist in literature, but the ones I'm presenting are sound, have a good track record, and can be exciting. Although there are variations to the methods in this book, these approaches collectively represent some of the best techniques available.

STRATEGY FORMATS

For each one of the strategy formats, we'll look at the specific steps of the process, then explore an example to show how the technique can be used.

Brainstorming: The Bread-and-Butter Process

Few people interested in creative thinking have not heard of brainstorming. First popularized in the 1950s by Alex Osborn, this technique has rattled around numerous board rooms, staff meetings, workshops, and conferences. Many people, however, have experienced an anemic version of the method, so there have been mixed reviews regarding its use.

I have been in less-than-robust brainstorming sessions. Little attempt was made to set the mood and to assure the climate for creative thinking. The storming rules were violated numerous times, including many not-so-subtle verbal and nonverbal judgments. Some leaders didn't facilitate well, even filtering out ideas that didn't match their vested interests. Many participants only got as wild and bold in storming ideas as newborn kittens gingerly lapping milk. Overly dominant people hogged the time and intimidated participants.

Brainstorming is a simple process. But the method and the participants require guidance, finesse, care, and feeding. By setting the climate, accentuating the storming rules, and adhering to the steps, you are assured of a productive creative strategy. The steps will be woven into the first of two examples below:

Example: Case of the Ultimate Mousetrap: Attraction Phase

Since in Chapter 6 we were commissioned to develop the ultimate mousetrap, we'd better earn our money. In our analysis of the functions needed in killing mice, we chose three subsystems: attracting, capturing, and eliminating the pesky animals. Let's apply brainstorming to developing ideas for attracting mice. Here's the prompt and a generated list of ideas for solving the first subsystem:

Outcome desired: Since we are just getting started on this project, we want to take a first cut at putting ideas on the table. Later we might want to go for more esoteric possibilities.
Strategy format choice: Brainstorming.
Question prompt: How can we attract mice to our trap?
Storm:
1. Let the question prompt provide the focus.
2. Review the storming rules, then post the rules where they can be seen throughout the session.
3. If in a group, members shout out their ideas. A scribe gathers the thoughts as quickly as possible, so individuals can make room for further options that have begun to percolate in the subconscious. List the proposals on newsprint so the ideas are visible and act as stimulants for further thoughts. Continue for 20 to 30 minutes. If you are by yourself, simply write the ideas on a sheet of paper. You could stand at an easel to allow yourself to move while you are thinking. A variation would be to use a tape recorder to gather ideas.

Sample:

- Mini-packages of mice delicacies
- Mirrors
- Shape and feel of their natural habitat
- Mice designer colors
- Gourmet cheeses for discriminating mice
- Mice ardor odor
- Pictures of mice
- Sculpted mice decoys
- Find out what mice like to eat, then display it
- High-energy pellets
- Shape like doll house or floor plant so it'll be attractive to householders. Mice seem to like things humans like.

- Sounds of mice having fun
- Etc.

4. Move to the check phase, where you will select ideas for analysis, refinement, and implementation.

As a foundation storming tool, brainstorming can work well in a variety of groups. Here's another example. Customers are asked to speculate what they would like a product to do for them.

Example: Case of Computer Friendliness

A forward-looking computer development company, Starcom, knows that to get ahead it has to think ahead. To aid their peering into a crystal ball and to create their own future, they set up a series of meetings with customers. They bring a variety of customers together, from novices fearful of the computer but forced into using it to highly knowledgeable users who play with every tool and gadget that comes along.

Starcom knows that computers require users to "work and wait," that is, users produce commands or type words then wait for the computer to catch up to their thinking like a cat-and-mouse game (those mice seem to get everywhere!). The company places a mixture of customers in groups of five to ten, believing they will get rich feedback even though the knowledge level among the users varies greatly.

Outcome desired: Since Starcom is after the customer's wish list, the company is willing to take a first cut at computer users' hopes and dreams. Refinement can come later. The company also believes these groups it assembles aren't quite ready for more arational techniques.
Strategy format choice: Brainstorming.
Question prompt: If your dreams could come true, what do you wish your computer could do for you?
Storm:

- Allow for normal communication
- Respond to my verbal commands

- Let me make thumbnail sketches on it
- Accommodate to my specific thinking and working style
- Automatically fix the words I regularly misspell
- Get into selected files and prepare a contrast of actual vs. budgeted costs when I verbally tell it to
- Read my mind
- Quickly bring me a key paragraph in something I've written
- Do what those information centers in department stores do—tell me where to find a pink ballerina outfit for my daughter or how to locate a military style men's hair brush
- Give me direct access to travel information
- Let me play "what if" scenarios quickly
- Move my data from one graphic style to another so I can see which one best conveys my message
- Set up a conference call just by my telling it to
- Etc.

Discussion

Brainstorming is basically simple. Yet setting the mood and following the storming rules make it come alive. That's true whether you are doing personal brainstorming in your journal or getting together with colleagues to gather ideas.

If you use a team approach, try to have no more than five to eight participants in the group. Add spice by giving each person a few markers (watercolor so there is no bleed-through), having the group cluster around a couple of sheets of newsprint taped to the wall, and letting everyone get in the act of writing down ideas—and using sketches, symbols, and other nonverbal expressions, too.

Brainwriting: Verbal Storms Without Group Discussion

This technique had an audience on the other side of the Atlantic before getting much of a toehold in America. Brainwriting has been gaining in popularity in the United States, probably because it allows participants at an idea-generation session to dish up suggestions at their own pace, without feeling intimidated by more vocal or dominant people. The steps are included in an illustration:

Example: Case of the Challenged O-Rings

Several years have passed since the world bolted upright with the explosion of the spaceship *Challenger*, Flight Mission 51-L. A presidential commission concluded that the culprit was the design of the sealing mechanism in one of the rocket boosters. The seals failed to prevent burning gases from escaping. You could debate which failed most, the rubber O-ring seals or the system that continued the launch in spite of available knowledge about vulnerable O-ring effectiveness at lower temperatures.[1]

Focus in this illustration on the faulty design. Some of the ideas I'm using have already been suggested in the literature. Without getting bogged down in technical detail, however, I want to conjecture how brainwriting could have been used to generate both practical and not-so-tame proposals.

Imagine you were in a group of engineers called together after the tragedy to come up with new ideas to improve the sealing mechanism in the booster rockets. Before beginning a search for ideas, you would already have analyzed the many factors critical to sealing those joints. You would have studied, for instance:

- potential for damage or contamination during assembly of the booster segments
- motor pressure on the tongue-and-groove joint connection, which was supposed to be protected from gases by two O-rings, relatively cheap bands of rubber encircling each joint
- ability of the O-rings to spring back to full position once compressed
- properties of the putty and other insulation protecting the O-rings

- possibility of water freezing in the channels in which O-rings rested
- booster strain and flexing once the engines are turned on for flight.

If your team decided to storm ideas, here is a sample of how you could have used a creative thinking technique:

Outcome desired: Because the sealing mechanism worked previously and radical changes might cause unneeded delays in the space program, you decide to aim for more resourceful and evolutionary rather than revolutionary proposals.
Strategy format choice: Brainwriting.
Question prompt: How can the field joints of the assembled booster segments be sealed more securely to handle launch and flight conditions?
Storm:
1. Use the question prompt.
2. Follow storming rules, especially piggybacking and allowing the wild.
3. Give each person a form divided into fifteen or eighteen cells, five or six rows with three cells in each row. Put one blank form in the center of the table, for the first person finished with a row.
4. Each person writes three solution ideas on the first row of a form. Add higher creative power by using words, images, symbols, and sketches. Work without group discussion.

Sample: See Figure 3.

5. When finished with a row, individuals place the form in the center. They take another form and complete another row, thinking up new ideas. If stymied in their thinking, they look at other ideas on the sheet for stimulation and piggybacking.
6. Repeat until four to six rows have been completed. Take 20 to 30 minutes for the idea generation.
7. The group or the leader may evaluate the ideas during the check phase.

148 CREATIVE THINKING AND PROBLEM SOLVING

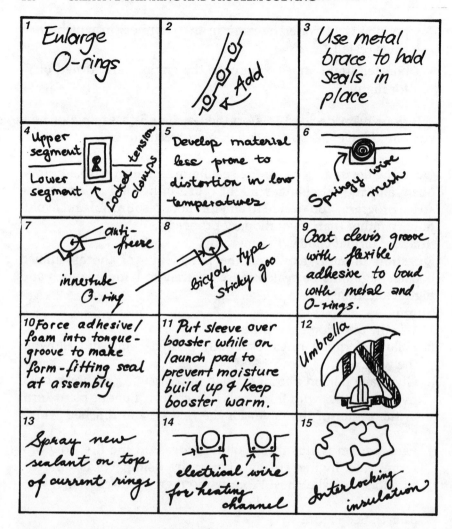

Figure 3. Example format of brainwriting.

Discussion

Many people like being able to go at their own pace in this technique, while also being in the presence of a group and seeing the development of other ideas. As the options get wilder and stretch the imagination more, there will often be chuckles from group members.

Another advantage of this process, especially when there are dominant individuals in the group, is the opportunity for introverts to shine. Extroverts are the ones who often get the group started during brainstorming. But they can intimidate the introverts.

When extroverts don't hear any ideas pouring out of the mouths of introverts, they believe that no one else has any ideas to contribute. The extroverts will then come up with even more ideas to fill the void! Brainwriting allows everyone to go at his or her own pace. Extroverts find out that other people have good ideas, too. As trust builds, the more verbal and interactive strategies may be used.

A half hour of generating ideas with either brainstorming or brainwriting, both lower power strategies, can produce 40 to 100-plus options. Of course, there are some duplicates, especially in brainwriting, since there is so much individual work. Don't worry that some of the ideas might not fit your problem perfectly or have to be discarded. Ideas often require further discussion, rework, and additional imaginative thought to make them more useful and feasible. Yet you have still gained a good range of ideas that can stretch your thinking and give you options with potential.

Mindmapping

Even if you are a diehard, plodding, stuck-in-the-groove, tenth generation rational thinker, there is hope for you. You can accentuate the arational part of you. You might even find you enjoy it! A place to start may very well be to create a map of your thoughts. Mindmapping is a good strategy to use for tapping intuition. This is a process that can be used by you alone or by you and your team.

Tony Buzan created the idea of a mindmap as a way of hooking thoughts together. His study of brain functions revealed that ideas often pour out of our minds in a nonlinear way. We need, he reasoned in his *Use Both Sides of Your Brain*, some way of linking thoughts without squeezing them into a less natural listing or step-by-step sequence.[2]

Gabriele Rico, who also delved heavily into a study of the modes of thought produced by the left and right hemispheres, produced a similar thought map. I like Phillip Goldberg's treatment of Rico's idea association clusters, presented in *The Intuitive Edge: Understanding and Developing Intuition*, as a complementary approach to Tony Buzan's method.[3]

Although many people take to mindmapping right away, others need to practice it a number of times—working on trusting their imagination and their ability to freely associate more openly each time they use the method. Free association doesn't work well from a hard focus.

Allow yourself, and your team if it's involved, to be relaxed, open, focusing softly, following the storming rules. Don't filter ideas or judge them or wonder what others would think if they saw them.

The steps will be explored as two examples, which are examined briefly below. In the first illustration, you'll see the Buzan format. Rico's method will be used in the second example.

Example: Case of the Effective Teamwork

A work unit meets at a retreat center for a team building session. Early on the first day, they realize they haven't had a clear idea of how they want to function as a team. They decide to take time to storm elements they should consider if they are to build a high performing team. The ideas they create with the mindmap form the basis for the afternoon's work, forging a vision and norms for how they want to operate together. They have fun adding symbols and sketches, making the mindmap more vivid.

Outcome desired: The team doesn't want anything fancy, just a chance to have an imaginative survey of their thoughts.
Strategy format choice: Mindmapping, the Buzan style.
Question prompt: Rather than a prompt, the group lets the word "teamwork" be their trigger.
Storm:
1. Choose a word or problem phrase. Put it into a nucleus bubble in the center of a sheet of newsprint.
2. Create associative spinoffs as ideas cluster and trigger other thoughts.
3. Collect themes or idea associations on paper.

a. Buzan uses spines and branches. Keep the lines curved to stay away from the straight lines of linear thought. Use terse, one to three word statements. Let the map build naturally. You could use different colors for the branches. Make it more graphic by sketching in stick figures, symbols, and nonverbal thoughts.

Sample: See Figure 4.

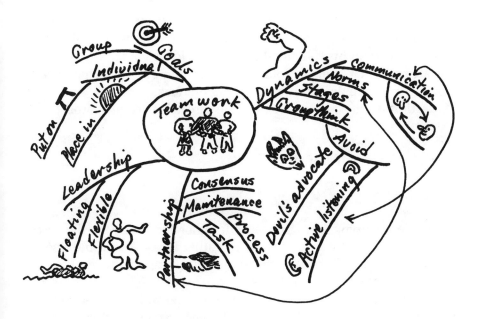

Figure 4. Buzan's style of mindmapping.

b. If you note that, in developing one branch or cluster, your current thought connects with another branch, draw an arrow between the thoughts. But keep on flowing. Don't get caught in analyzing or judging. Hold that off for a while and your intuition will likely deliver some new insights to illuminate or solve the issue in the nucleus of the map.

4. When the associations are wrung out, go to the check phase.

152   CREATIVE THINKING AND PROBLEM SOLVING

Example: Case of the Questioned Career

A behavioral scientist experiences a midlife crisis. She is in turmoil over what career road to take. One Saturday she plunks down in a lounge chair on the patio in her backyard. She decides to storm some ideas about her career, writing the ideas in the journal resting on her lap. Although the storming takes less than 10 minutes, she sits on the patio for more than an hour, reflecting on the mental map she has created.

Outcome desired: She's interested in mapping the random thoughts entering her mind, letting free association and intuition guide her thoughts. She'll take whatever ideas come.
Strategy format choice: Goldberg's rendition of Rico, which uses arrows and bubbles for the clustering of ideas.
Question prompt: What are possible directions for my career?
Storm:

Sample: Examine Figure 5.

Discussion

If mindmapping in a group, with either the Buzan or the Rico methods, there are several ways to gather ideas. You could use a facilitator, letting her shape the map while group members suggest where to hook ideas.

Or put a big sheet of butcher paper on the wall. Give everyone in the group four or five colored markers. Have them stand around the paper, jointly building the mindmap. Individuals may announce what they are adding, but refrain from discussion. Much energy and synergy can be created with this group exercise. Or have individuals create individual maps on newsprint. Participants are invited to share theirs with the group. A composite mindmap could even be created.

If conducting mindmapping in a group, 20 to 35 minutes may be sufficient time to complete a map. If you do this alone, your pace may vary from the group's speed. While the time may be approximately the same, go by your inner clock. Once you have learned to let the mindmap freely run its course, your intuition

Figure 5. Rico's style of mindmapping.

will tell you when you have collected sufficient material. You'll be able to add more later if you train yourself to keep your mind open for additional insights.

Analogy Storm

To create an analogy storm, you let the target lie fallow for a while. During that time, you take a 1- to 2-hour incubation and analogy journey, moving through a series of analogies that seemingly take you away from the problem or opportunity.

While analogies have been used since the beginning of humankind, two men, William Gordon and George Prince, set the use of analogies into a helpful creative thinking process called *synectics*. Gordon's *Synectics* and Prince's *The Practice of Creativity* are gold mines for gathering further precious material about the use of analogies in creative thinking. The synectics method calls for a leader or facilitator who is skilled enough to guide a group through its nuances.[4,5] Analogy storm builds on their seminal ideas. Here is what analogy storm asks you to do:

- Take a journey into the land of analogies, a world of word pictures that resemble in some form or characteristic something you know. Be distracted from your original problem. You were developing a mind-set. Take time to unlock yourself from the mind-sets you were developing about the situation and the target.
- During the trip, explore and play with analogies. Travel like a curious, relaxed tourist. Poke around with metaphors. Get excited by the new things you see.
- If you are to make it to the end of the journey, you have to keep moving. A few characteristics will begin to stand out as memorable at each new stop on your travels. Let the journey be refreshing, so that, once back at the old problem, you can begin to make new connections and see things in a new light.

There is a rhythm or groove that you often have to get into to capture the power of metaphoric thinking. You follow a train of different types of analogies that build on each other until you gain unique viewpoints and inventive ideas for meeting the wishes or conditions you set up. The analogy storm technique simply structures the flow of analogies and the plucking of characteristics or attributes that can stimulate imaginative ideas.

Many people jump right in to take this analogy journey. Others find themselves resisting somewhat. If your mind struggles with analogy storm, demanding that you be more logical and practical and have an immediate payoff, try these tips:

- Tell yourself analogies have their own logic and beauty.
- Give up precision and the concrete and take on approximation for a while.
- Play with more relaxation, spontaneity, and softer focus.

- Enjoy being bombarded with analogies for a set period of time, one half to one hour, for example. Remember that the intent is not to milk all analogies, but to let them distract your hard focus and perhaps jar or transport you into seeing things from fresh perspectives.
- Give yourself permission to pick up an analogy you like, extract an interesting characteristic from it, and move on.

To provide special guidance through the analogy storm, we'll explore a case, moving step-by-step and looking at a brief illustration with each step as needed.

Example: Case of the Ultimate Mousetrap: Capture Phase

Let's garner another set of ideas to improve the proverbial mousetrap. We used that timeworn target to help us generate proposals to attract mice with brainstorming as a strategy. We can get good mileage and a different perspective by applying analogy storming to the second subsystem, "capturing mice."

By this time you might have caught on to the fact that our target may be a bit too narrow. In the brainstorming example, we sought ways of getting mice to come to us so we could do something to them. Certainly we could refine the target to include ways to get rid of mice without necessarily attracting or capturing them. Dumping a boxcar full of cats in our living room or throwing a canister of nerve gas in the crawl space under our home might cut out some of the mice population. For the sake of working within a few constraints, however, let's stick to the notion of "capturing mice" as the second subsystem where we need ideas.

Have some fun using this simple, homey illustration to learn more about the use of metaphors in thinking creatively. Analogy storming can be enjoyable and energizing, but does take effort.

Outcome desired: You want to have the boundaries of your thinking pushed in new ways.
Strategy format choice: Analogy storming.
Question prompt: Incorporated as wishes in the first step below.

Storm:

1. Create fantasy analogies or wishes to dream about the desired state. These are akin to a series of target statements. You could start the description with the words, "I wish ... ," then add the condition you would like to see exist.

   Sample: I wish I had a sure-fire way of capturing mice so we could then eliminate them.

2. Choose an analogy world.

   Sample: Try the world of nature, for it is an abundant field in which to let your imagination romp and to pluck metaphors.

3. Develop direct analogy links between the wish (target) and the world of nature.

   Sample: Your facilitator asks: "What are five to ten ways that capturing takes place in the world of nature?" The list your group develops is:

   - Octopus wrapping tentacles around something
   - Whales straining meals through horny baleen plates
   - Anteater flicking out tongue
   - Spider spinning a web
   - Lobster grabbing with claws
   - Snake wrapping itself around a body
   - Cave man slugging a woman over the head and dragging her to a cave
   - Cat sneaking up on bird, then pouncing on it
   - Amoeba flowing around food
   - Venus plant trapping flies that wander into its mouth

4. Choose an intriguing or exciting analogy from the direct analogy list. You and your co-group members identify with that analogy, emotionally and enthusiastically entering into the life of that metaphoric object or process. Individuals describe what it feels like to be or to have the characteristics or mechanisms of the chosen personal analogy. The descriptive phrases are listed on newsprint.

   Sample: The elegance of the spider spinning a web captures your group's attention as an intriguing direct analogy. The facilitator begins the task with, "Identify with the spider. Give a phrase or sentence to describe what it feels like to be a spider creating a web." Selected answers:

   - Frenetic weaver
   - Daring adventurer
   - Exotic creator
   - Delicate sensitivity to motion
   - Extending self
   - Sneaky and sly
   - Creating mystery and beauty

5. From that list of descriptive phrases, choose an analogy that intrigues you, one that is a good description of the personal analogy. This time create paradox, seeking analogies that are a contradictory description of the chosen personal analogy.

   Sample: "Exotic creator" is tapped by your team as a quick-witted, terse descriptor of the personal analogy. Then the facilitator asks, "What's a contradictory analogy to 'exotic creator'? What metaphors contrast or contradict or apply a clever, opposite, or paradoxical twist to the phrase we've chosen?" Examples of these new analogies:

- Slumlord
- Flea marketeer
- Messy scribbler
- Colorless smasher
- Native terminator
- Creepy pusher
- Couch potato
- Messy manipulator

6. Choose a contradictory or paradoxical analogy for further development. If the original problem was technical, attempt to get into technical or natural worlds you can mine for metaphors. This time you look for five to ten direct analogies.

   Sample: "Creepy pusher" catches the eye of your group. The facilitator begins to move the team closer to specifics by asking, "What direct analogies from the world of technology come to mind when you think of 'creepy pusher'?" Some direct analogies that are generated:

   - Motorized hydraulic lift moving crates around a warehouse
   - Bottles being filled as they creep along in a bottling plant
   - An antidepressant
   - Cocaine
   - Arteries and heart probed with a catheter

7. From the list of direct analogies you generate, apply the force fit concept.
   a. Using the systems, processes, and mechanisms in this new list, you storm solution possibilities. New connections are coerced from the direct analogies. It is in the

excursion away from the problem, while your mind is incubating and making free associations, that new ways of looking at the issues and solutions emerge.

b. You are now at the point of generating bright ideas to solve your original problem. You traveled into the world of analogies to break any mind-sets that formed while analyzing the situation and identifying targets. Incubation has been taking place, with your intuitive and preconscious thought processes working out of sight like busy beavers.

Sample: The facilitator says: "You've gone on an excursion away from the original target or wish. Now create connections between the list you just generated and your desire for ways to capture mice. Force connections to provide proposals." A few suggestions:

- Probe down a mouse hole with a catheter to find mice, then grab them with some mechanism from the first set of direct analogies
- Put "uppers" in pellets that make the mice get high very quickly, so fast they are dazzled and immobilized
- As mice enter a track, have them sucked into a container
- Send a snake down the hole
- Have a robot move to different locations—grabbing the mice carefully, securely

8. Discuss how these ideas could be shaped into doable, useful options.

Sample: Your facilitator asks, "Do we need to work these ideas any more before we can evaluate how well they fit our requirements for coming up with a good mechanism for the capture phase of developing the ultimate mousetrap?" The group suggests they would like to explore the "uppers" idea more. They storm some possible ways to use that idea. Several thoughts:

- Develop substances that would paralyze mice when they eat
- Mix the immobilizer with food that's attractive to mice
- Have containers mice would enter to eat but not get out quickly

9. Go to the check phase to assess these proposals.

Evaluation takes place in a number of the steps of analogy storm. You are asked to storm then choose analogies, storm then choose analogies. Even at the end of the journey into the land of analogies, you may have to move back and forth between storming ideas and making choices. Solution proposals can become novel, deliriously funny, boundary challenging. During analogy storm, you may need to help tame and use the unusual approaches.

The same may happen in the check phase, where greater clarification of ideas is often sought. You keep looking for the most promising brainchild, but sometimes, when playing with novel ideas, you have to think of specific ways to help unusual ideas become winners.

Informal Use

Of course, you can use analogies without using the more formal process we just discussed. You may want to weave the creative power of analogies into your thinking just to stretch and stimulate your imagination.

Search for analogies to some condition or target. There are a host of arenas that are full of metaphors. Choose a world—the sphere of medicine, sports, nature, business, insect life, outer space, animals, ghosts, clothing, mining, plants, games, shapes, shades, leisure, knitting, computers, relationships, pumping iron, baby products, water crafts, fruit, rocks, or whatever your mind can conjure.

Create a storm of analogies from a world that lets you make associations. Make connections from the metaphorical elements back to the condition or target. Natural worlds are often fruitful for pulling out analogies, for they contain potent processes and life cycles.

Rather than use the train of analogies, as in the synectics approach, see how you can be stimulated by using just one type of analogy. For instance, it can be helpful to use direct metaphors related to mechanisms or characteristics in the problem you are exploring.

In trying to find an analogy to characterize your research department, you could take a page from the world of sports. "The staff in the Research and Development section in our company is like a group of competing gymnasts. We have many members of the staff trying spectacular individual feats—speeding toward a single goal, doing gravity-defying twists in the air, hoping to land on their feet brilliantly so they can attract points, glory, and further financial support. Yet they are like a poor team, rife with dissension, not appreciating the efforts of their colleagues, not inspiring each other with their energy, skill, and team spirit."

Discussion

Proposals at the end of the search stage often require further work, as is true of most ideas. Analogy provides approximation and new angles for looking at things. Refinement of a brainchild may take evaluation and polishing, even additional creative thinking to get the kinks out of an idea. Participants in analogy storming often come up with new thoughts about the hoped-for outcomes. They also gain insights into the mechanisms that could be employed in coming up with a final fix or outcome.

Picture Tour

Just as analogies can trigger creative thinking, pictures can also be used as a springboard toward innovation. We'll weave a case and steps together, to "get up close and personal" with this useful method.

Example: Case of the Parched Land

Let's go back to our mythical western state, DesertSun. It is scorching in the heat of the day and struggling with its overtaxed water supply. New thinking is needed to spur renewed efforts to find proposals that could satisfy the thirst of homeowners, industry, and farms.

Outcome desired: The governor of DesertSun asked for inventive thinking and expanded approaches. Bold reasoning calls for the arational.
Strategy format choice: Picture tour.
Question prompt: How can we have water more available for home, farm, and industrial use?
Storm:
1. After developing the targets and settling in on where you want to aim your search, have a 4- to 5-minute idea storming splurge.

   Sample:

   - Allow home sprinkling of gardens and lawns only on certain days
   - Have the city fine violators
   - Develop and use more drought-resistant crops
   - Insist everyone put a brick in the water closet of their toilets
   - Develop a poster campaign to publicize conservation
   - Create toilets with more flush power and less water

2. Redo the target statement, if needed, to handle any new understandings of the situation arising from the idea splurge.
3. Take a journey away from the situation by viewing slides on a screen. Tour scenes of action, people, and scenery. Extract a characteristic from each slide, one that intrigues you or is an essential feature of the slide. Participants silently develop their own lists on sheets of paper. Use nine to eleven slides. Observe each slide for 20 to 30 seconds. Have relaxing mood music playing in the background.

4. A facilitator posts on newsprint one or two characteristics that are most interesting to each member. Gather elements from one, a few, or all the pictures.
   a. If someone in the group sponsored the problem in the first place, that person could choose one to three features from the list of characteristics that piques her interest. Or the team could choose an element or two as a thought trigger. Or let all the items on the newsprint be stimulators.
   b. Use those selected characteristics as stimuli to shape or force fit ideas that might help the target become reality. Observe storming rules.

   Sample: See Table 1.

5. Move to the check phase.

Discussion

Through research at Battelle Memorial Institute's laboratories in Europe, Horst Geschka, Ute von Reibnitz, and Kjetil Storvik found that picture tour, what they called *picture confrontation,* was as effective as the synectics method in producing useful, novel ideas.[6]

A wrinkle they sometimes used in place of projecting slide images on the screen was to provide individual participants with packets of pictures. During alone time, individuals would peruse the images and extract intriguing characteristics. The rest of the session proceeded basically the same as in picture tour.

Pictures for the visual journey preferably should not be connected to the situation targeted for the search. Nor should they be too emotionally gripping. Les Fehmi, the open focus researcher, believes that pictures that have depth provide unique visual stimulation. Where possible, therefore, use pictures that provide a three-dimensional perspective.

For more on the use of pictures and hand-drawn imagery, see *The Wholeschool Book: Teaching and Learning Late in the 20th Century* by Bob Samples, Cheryl Charles, and Dick Barnhart. Also read Robert McKim's *Thinking Visually.*

Table 1. Ideas Triggered by Characteristics Gleaned from a "Tour" of Selected Pictures

Pictures	Fascinating Characteristics	Force Fit Ideas
New trees coming out of decaying host tree in rain forest	Nature's cycle	Use anaerobic process in toilets
		Develop water purification system for each home
Billowy storm clouds hovering over mountain peaks	Cloud turbulence	Tie numerous hot air balloons together to simulate air turbulence
	Peaks cutting clouds	Create mountains of garbage near cities for weather stimulation
Closeup of woman in crowd whose features are mostly covered by a wide-brimmed hat	Covered mystery	Hide new houses underground to control water loss
Tourist observing Michelangelo's sculpture of David	Reshaped material	Burn tires. Condense plume of smoke. Make water.
		Import huge chunks of ice from Alaska by barge. Wrap in thin Styrofoam. Ship by rail to DesertSun.
		Reshape water resources with major conservation effort and government inducements.
Tumbleweeds tossed around by wind	Bouncing weeds	Develop sponge baths. Have sponge balls containing measured water blown at you.

Imaging the Future

This is another creative thinking strategy of the higher imaginative power category. It is a technique that stimulates the imagination by putting the future in a crystal ball—and having a ball creating vision. Peering into or speculating about the future has an exciting, sometimes emotional, edge to it. Science fiction writers, for instance, have a field day tantalizing us with scenes and experiences about the future.

In not quite so dramatic a way, imaging the future provides an opportunity to set aside the struggles of the day and jump into a future time, to dream, speculate, and pose new possibilities. The leap into the future may be a few months or a few years. But the strategy gives you an opportunity to create something new, rather than get bogged down in the constraints of the present.

Robert Fox, Ron Lippitt, and Eva Schindler-Rainman developed the framework for this strategy. Their *Towards a Humane Society: Images of Potentiality* presents the background for this technique.[7] I have often used the process with teams or individuals in organizational settings where strategic or futuristic thinking is wanted.

This method is not for making pure pie in the sky. It can, for instance, provide sound proposals for creating optimal conditions for siting a new facility, a five-year program for a research team, or key staff roles for keeping an organization at a cutting edge. This strategy is also excellent when used during the target phase to create imaginative outcomes, targets, or conditions for a desired state.

There are several steps to this process. We'll look at them as we explore an illustration.

Example: Case of the Flourishing Environment

Imagine that you are in a company with good people, but with a work climate that has been tuned more to Attila the Hun than a challenging and entrepreneurial spirit. A new chief executive officer has established a task force to see what can be done to stimulate the environment and allow imaginative breezes to blow through the company. You and your task force members meet to come up with ideas that will promote a new vision.

Outcome desired: You want to rattle mind-sets, stirring yourselves and posing new directions.
Strategy format choice: Imaging the future.
Question prompt: See step 3 below.
Storm:
1. Decide to stretch out your thinking into the future, preferably at least several months away. You may want to focus on a range of time, one to two years, for instance.
2. Create an imaginary vantage point in the sky—a cloud, magic carpet, helicopter, or balloon. Give yourself the power to look down from your unique viewpoint and survey all that is taking place below you. Your eyes can pierce buildings, like Superman or Superwoman, to observe people at work or play. You can stand back to see the big picture or move in for a microscopic view. Your ears eavesdrop on conversations. You are able to see yourself or co-workers engaged in activities.
3. Use a question prompt that will focus on a hoped for outcome. For example, "What are the exciting, meaningful things being done by our work group one year from now?" Or put your attention on roles undertaken in the future. For instance, "What are the fascinating things we see ourselves doing to pull off the outcomes we desire?"

   Sample: What are exciting, meaningful things we see ourselves doing in our organization to stimulate and nourish creative thinking?

4. Create a storm of mental images.
   a. When thinking of events in the past, present, or future, you usually have a mental picture of what is happening. This imaging is usually done so naturally you may allow the image to be blurred or not fully in consciousness. In this strategy, sharpen the image as if you were adjusting the focus on a video camera. If you think of yourself being involved in an exciting project in the future, make the mental image of yourself engaging in that project as vivid as

possible. Summarize the image into a concise statement that can be posted on newsprint by a scribe.

   b. You can restrict yourself to images that are doable and practical. Or you can add spice, allowing wild, far-out images, as well. The relish of the latter can make even the practical richer.
5. A facilitator collects the images on newsprint. Thirty minutes is an average amount of time to allow for the storming.

Sample:

- People encouraged to exceed boundaries of responsibility
- An improved employee lounge
- Swapping ideas with other departments
- Slush fund for small projects
- Fish tanks with guppies in the office
- Can check out Walk-mans with stimulating music
- Creativity computer network set up
- Corporate culture reset
- Quiet places for incubation
- Executive dining room eliminated
- Corporate creed regarding creativity etched on a plaque
- Ideas sought from customers and vendors
- Creative thinking classes taught on a regular basis
- One restaurant meal per month paid for so employees can talk about new ideas with co-workers or management
- An extravagance or costume day
- Meditation techniques introduced
- Executives first get, then give creativity courses
- Isolation periods with no phones or interruptions
- Healthy diet for good thinking recommended

- Calisthenics every morning
- Atrium type of open atmosphere
- Guru hired
- Another company invited to cross-fertilization sessions
- Off-site creativity sessions to handle complex issues
- Romper room
- Budget for creative thinking sessions and backing innovative ideas
- Ability to laugh at oneself encouraged
- Outside stimuli, personalities, etc. used
- Freedom to "screw up"
- Seances
- Increased employee confidence
- Meetings in the spa
- Rotating creativity council

6. Go to the check phase, where you choose the most exciting images. Then analyze the best images to determine what might inhibit or facilitate each choice outcome or role.

Discussion

The above image storm is a collection of ideas actually created by my workshop attendees when they were presented with a problem similar to the case of the flourishing environment. These are only a few of the many ideas they produced. I held out some of the wild images: paper cup fights, sex therapists, sushi for lunch, more wine breaks, staff meetings at a go-go, free funerals. These slightly off-beat proposals helped pump up enthusiasm and imagination so further fertile and useful images could emerge.

Ron Lippitt, one of the social science deans in America, used his futuristic thinking processes with large community and educational groups. As a prelude to visioning, he would often have a big gathering of people, broken into smaller work teams, preview "prouds" and "sorrys." *Prouds* were the accomplishments of their wider community that make group members feel good. *Sorrys* were the issues and lack of achievement in specific areas that cause members to yearn for solutions. A technical team could also begin a period of thinking about the future by spending one to two hours on their prouds and sorrys, then conducting "imaging the future."

□ □ □

Higher power storming strategies have more breakthrough energy. They endeavor to tap the more imaginatively vigorous arational modes and tend to get participants more emotionally involved. Since there is a combination of thinking modalities, these strategies have a greater probability of pushing into new territory and producing more original options.

All the strategy formats are useful, however. You have to pick and choose to meet your comfort level and the needs of the situation. Table 2 presents a recap of the creative thinking processes presented in this chapter, as a handy guide for what they are and can do. There are pluses and minuses to consider.

Your decision about which technique to use may revolve around the degree of novelty you want, your comfort level, how tolerant colleagues might be of the arational, and the skill needed to carry out the process. For instance, lower creative power techniques—brainstorming, brainwriting and mindmapping—may be aided by a facilitator to keep a group on track, provide an atmosphere of neutrality and freedom, and, in some cases, collect ideas on newsprint.

Higher power creative processes, except for such personal techniques as meditation and dream work, take a better understanding of the subtleties of the methods. Facilitators tuned to arational methods can be useful in guiding people through the more numerous steps and the bolder perspectives of analogy storm, picture tour, and imaging the future.

Table 2. Summary of Six Creative Thinking Strategy Formats

Strategy Format	Description
Brainstorming	Making ideas visible as they come to mind. Traditionally given verbally in a group and recorded on newsprint by a scribe. Sound, quick, potent if storming rules are honored. Can be curtailed if strong person dominates. Can be done solo. Enhanced when symbols, images, metaphors are used.
Brainwriting	Eliciting ideas without group discussion. Participants write or draw thoughts on a form with rows of boxes. Papers are switched when a row is finished. Storming rules are still important, especially piggybacking and bold thinking. Good for keeping individual ideas flowing, nontalkers contributing. Some extroverts miss the verbal stimulation.
Mindmapping	Capturing ideas in nonlinear fashion. Thoughts branch outward from a trigger word, each thought moving from central themes to subelements. Branches grow as intuition makes new connections. Can be blunted when thoughts are prejudged.
Analogy Storm	Searching for analogies that give insights into characteristics of a current or desired condition. The problem is left alone for a while to follow a series of analogies, potentially leading to fresh or novel viewpoints and solutions. In its full form, the technique needs a facilitator to guide the process. Takes an ability to play with arational media.
Picture Tour	Using pictures as stimuli to break mind-sets. Characteristics of pictures are identified, then used as bridges to shape solutions. Requires an appreciation of the visual and the force fitting of neutral elements of pictures into solution possibilities.
Imaging the Future	Conjuring up mental images full of future potential. List exciting outcomes you visualize happening at some designated future time period. Those who can only brainstorm rather than have fun creating mental images can still find this method fruitful.

COMMUNICATION TIPS POSTSCRIPT

Crucial to the free flow of ideas is open communication. If you are operating in a group setting, there are things you can do to assure a stream of good proposals. You have to build, of course, on your communication efforts undertaken during the target phase. Here are a few more things you can do in the search phase:

- **Set tone.** Set a good tone for the generation of ideas. Have a mini-lecture on creative thinking. Briefly describe how locked in we can get to business and reasoning as usual, pile-driving ahead caught in our mind-sets. Tell the importance of pushing the boundaries of our thinking. Give encouragement in how we need to try new ways of thinking, even if we are a bit uncomfortable at first.

  List on newsprint and describe the storming rules. Even when people know what they are, it is helpful to reestablish the norms for the whole group. When, in the heat of the storm, someone becomes narrow-focused or judgmental, it is easy to remind that person that the group has agreed on a more open approach. Support from the manager for the need to be imaginative and to follow the storming guidelines will underscore the mood that is needed for the search phase.

- **Facilitate, don't filter.** The person who is writing down the ideas on newsprint must treat each option that is generated as a bona fide proposal. Any judging by the scribe about what gets listed and what doesn't will be carefully scrutinized by the idea-generators. Some paraphrasing may be useful, but collecting participant's actual words while ideas are coming like popping corn shows that the facilitator means business in honoring ideas. To show the integrity of the scribe, even funny, off-hand comments can be posted.

- **Use caution in showing crazy ideas, strange analogies, or stormed images to those outside the circle of folks that generated them.** Since you want freewheeling, you need to post the wild proposals as idea teasers or uncut gems. Those who were not a part of the storming session, however, may not appreciate the context or technique in which the ideas were spawned. The data should belong to the group that hatched the ideas until evaluation takes place and you refine the proposals for action.

- **Make ample use of a journal.** I have mentioned journals before as being one of the most vital tools available to you for capturing and displaying ideas. Their use provides a way for you to communicate with yourself. By letting your hand be a conduit for intuitive, preconscious thoughts, you have a channel for expression of your imagination.

A journal can be used to collect thoughts throughout the creative cycle. But I want to underscore how a journal also lets you develop a record which can be vital if you try to patent an idea. Here are some suggestions that can help your recordkeeping:

- Use a bound book, rather than a loose-leaf notebook. Pages torn from a book leave tell-tale traces. Patent lawyers get heartburn if they can't establish ironclad continuity of thought.
- Establish the journal as yours from page 1, with your name and the date you started your journey of ideas.
- Apply discipline. For each entry, start with the date. Show your train of thought, including where you got any part of your idea—whether from a dream, a colleague, a scholarly paper, a talk, an experiment, or even while being absorbed in a video thriller. Use permanent ink if you want the most indelible record. I have mixed feelings about being constrained to just permanent ink, however. I believe that using color in the journal can help stimulate thought, and many felt-tipped markers are watercolor.
- Sketch and label your drawings. If possible, describe or show how your proposal makes a difference.
- If you make a journal entry you believe is creative, go to someone who can officially witness your proposal. Companies often have a representative of their legal staff verify he has "read and understood" the proposal, with you sign-

> **XEROX CORPORATION**
>
> DATE Mar 22, 1965 SUBJECT Induction Imaging PROJECT NO.
>
> Because magnetic brush development is capable of extremely rapid response to fields of very short duration, we have demonstrated its use with conventional interposition development. We have, in fact, compared results by this technique with those of induction imaging at various relative humidities. Interposition development is done by exposing (Se/µSe plate) to a negative original, taping one end of paper to the plate and pressing it into contact using a biased (+360 volts) magnetic brush as shown: 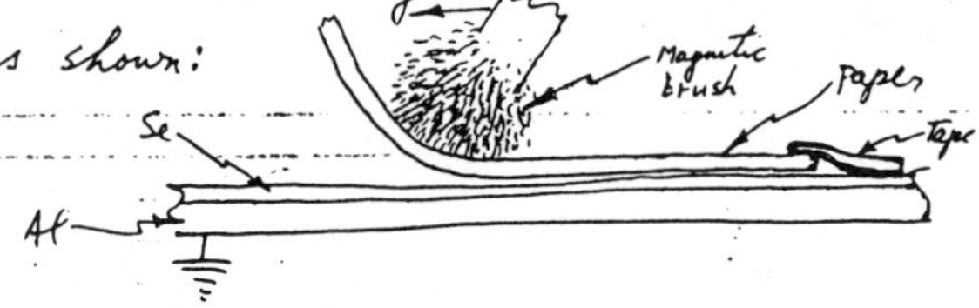
>
> [sketch: Magnetic brush, Paper, Tape, Se, Al]
>
> The same brush is then moved over to the other half of the plate, which was exposed to a positive image, and moved from left to right with a bias of +120 volts while the paper is lifted from the xerographic plate, to give develop the induced image. The next page shows samples by both methods at various humidities.
>
> WITNESSED AND UNDERSTOOD
> SIGNED Peter R. Keenan
> DATE March 26, 1965
>
> WORK AND RECORD
> SIGNED Rud Gundlach
> DATE Mar 22, 1965

Figure 6. Project notes of a Bob Gundlach idea for a xerographic process. From *Inventors at Work*. Reprinted by permission of Microsoft Press. Copyright 1988 by Microsoft Press. All rights reserved.

ing and dating the entry. As an example, see Figure 6 for Bob Gundlach's project notes, with a verification of one of his xerographic ideas.

SHIFTING GEARS

Once we are overflowing with options, we need to move to the check phase. There's some hard work and more creative effort ahead as we continue the journey through the creative cycle. Our task will be to make choices.

8   Check: Assessing Options

Innovation requires a good idea, initiative and a few friends.
—HERBERT SHEPARD

Artichokes used to annoy me, until I got to thinking of their leaves as petticoats.
—OLIVER WENDELL HOLMES

One of the inventors of the day is the comic strip character Wizard of Id. Whether trying to invent something or searching for a solution, the wizard often narrows his eyes and points at the vat in his dungeon, sometimes muttering a few magical words. A bolt of energy and a *kazam* rattles the vat, and he has a chance to see the results of his creating. His invention, for better or worse, is then out in the open for evaluation. Sometimes his concoctions shine. Sometimes they fizzle. At other times they come out distorted but unique.

When we are pressed to the wall by a special situation and need to come up with unusual ideas in a hurry, we might like a bit of the wizard's magical power. In a storming-calming-accessing session, we focus mental and imaginative energy, concocting ideas. Tapping our intuition and capitalizing on our free associations often produce surprises and awe. Creating can put us in a magical mood.

Some of our ideas may shine brilliantly. Others may fade. Still others may take unusual shapes. Like the concoctions of the Wizard of Id, our mental creations must be evaluated and choices need to be made—by ourselves, sponsors, or those with other perspectives. During this assessment period, it is important to maintain the mood of wonder and exploration begun when we generated the options.

PULL OUT THE MEASURING TAPE

As a result of searching for ideas, you now have a pile of proposals in front of you. It's time to sort through that pile, evaluating which suggestions best meet your desired target and analyzing how well your chosen ideas will sail when implemented. You *check the ideas for usefulness and potential*. In this phase, you *match proposals against criteria and size up any potential problems that may be inherent in the options*.

Your imagination may have given birth to many ideas. Your mind may have conjured up a mousetrap that vaporizes varmints, a computer that corrects your mistakes, an exotic company hideaway for fostering creative thought, a mountain of garbage to stimulate a change in weather patterns, a sleeve spacecraft could wear on the launch pad as protection from rain and cold, a drought-resistant fruit, a design for an automobile-sized Hovercraft, a formula for devouring oil spills, a pocket-sized device to test the cholesterol content of food.

Ideas such as these are impressions, opinions, images, or notions. To be fully useful, the proposals often need further analysis, testing, or reworking. Ideas often point in a promising direction, but require study to determine their full potential.

Your immediate concern at this point in the creative cycle is to sift through ideas you generated to find those with the best potential for meeting your target. During the check phase, you first make sure you understand the intention of each idea suggestion. Then you begin to make the list manageable, sorting and winnowing ideas. The most promising proposals are held up for scrutiny, like a winemaker testing wine. He holds a glass of new wine toward the light to assess its color and clarity, then swishes the wine before sniffing its aroma. The expert finally examines its taste with a mouth cleansed after each sip.

Although you need to use keen judgment like the winemaker, you need to take another step or two. Some of your ideas may require another round of creative thinking to polish them for further evaluation. Once you have chosen an option, you'll need to sweep the proposal for mines that might blow the idea out of the water. The steps of this process of studying new ideas are depicted in Figure 1.

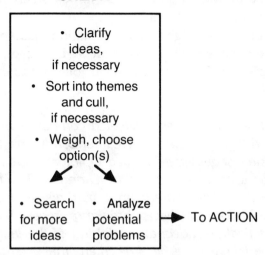

Figure 1. The check process.

The complexity of proposals you generate and the context for having them bear fruit can vary considerably. Therefore, techniques described in this chapter must be tailored to meet your needs. Many of the methods described here can be combined for a more elaborate assessment. Several of the techniques will be illustrated.

You may have other evaluation methods that will serve you well. Since the emphasis in this book is on practical steps you can take alone or in a team meeting, statistical models are not being used.

Design your evaluation process by determining how much sifting and screening you'll need. If you are sifting dirt from an archaeological digging, you'll probably use a screen with a fine mesh to prevent small artifacts from being lost. If you are laying a foundation of gravel for your new sidewalk, you'll use a screen large enough to allow small- to moderate-sized rocks to slip through.

This is decisionmaking time, of course. If many of your ideas will meet your goal and the costs aren't too high, you may not need elaborate screening to make an adequate decision. If the impacts would rattle cages, shake foundations, and put a gigantic hole in a bundle of money, you'll most likely need a more thorough evaluation process.

Enormously high stake evaluations can pose distortion in judgment. When Flight Mission 51-L was on the launch pad, thousands of pieces of data were collected to evaluate the readiness of the *Challenger* for orbit.

Massive, highly sophisticated computers monitored the sensors and flight equipment. Weather conditions, temperature readings, and test data were studied. Opinions were weighed. Information about the sensitivity of the O-rings to low temperatures was disregarded.[1] The countdown toward disaster continued because the evaluation and decisionmaking process broke down. There was tragically inadequate weighing of the factors. Political and managerial pressures became criteria that overrode technical data.

Most evaluations you conduct don't carry the potential wallop of a space flight. But it is still important to decide how many and what kinds of screens you need to arrive at a sound selection of options. Figure 2 provides an overview of the screens that can

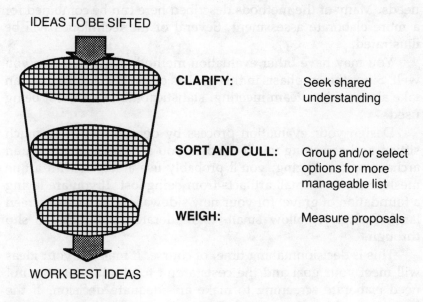

Figure 2. Evaluation screens.

be used for sifting through the newly generated proposals—clarifying, sorting and culling, and weighing.

You could stop at any screen, then make a decision. Be cautious about tossing ideas away too rapidly. Many winning ideas go unnoticed or are thrown away because evaluators don't fully understand them. Going through successive screens would add more rigor, clarity, and information as you push for the most objective data on which to make a decision.

In the WIZARD OF ID, measurements are comic by nature. At times the wizard is his own judge. The pint-sized king, his sponsor and benefactor, often evaluates the inventor of Id in bold terms—adulation or a stint in the dungeon. The wizard's wife, one of his major foils, sometimes assesses the value of his handiwork. The cartoon shows her being the brunt of one of his "inventions."

Reprinted by permission of Johnny Hart and NAS Inc.

Evaluation of your creative options may require a variety of perspectives. Once you have teased ideas out of the mind and onto paper, you can proceed with the assessment privately. Or you may engage your sponsor, team, or client in the process. You could call upon an expert, facilitator, or leader to carry out some of the evaluation tasks. A subgroup might also be used.

Who should you involve in the evaluating and selecting? If you want sage judgment and involvement, your decision may boil down to how you answer two questions:

- **Do you want others to own the chosen proposals and help implement them?** If so, the people who are going to carry out the work should participate. Sponsors and others who have to provide resources and nurture an idea through a system need to get their fingerprints on the options somewhere, even if only in helping determine the forces that will aid or retard the development of the idea.

- **Do you have the knowledge to fully evaluate the idea?** Even if you or your team have the technical expertise to measure the worth of an idea, a host of other considerations may make or break a proposal. You may want to go after the viewpoints of marketing, manufacturing, customers, stockholders, business sectors, and human factors to balance technical assessments.

An atmosphere of open discussion must prevail throughout the creative cycle. Even during this phase of making judgments, you want to create a continued soft focus mood in which consensus, cooperation, and collaboration mark any group interaction. Honoring all ideas, even zany ones that everyone knows are more fun than workable, means that you appreciate the thinking processes of your colleagues. To foster that honoring mood, follow the communication tips at the end of the chapter.

To help you decide what assessment screens to apply, I'll present a short description of each step in the sifting process. Choose evaluation and selection methods to meet your specific needs.

1. Clarify Ideas, if Necessary

> **If a group participated in the search for ideas, ask, "Before we evaluate and select the ideas, glance through the list. Are there any items that you would like to have clarified? Or are there proposals any of you gave that you would like to explain briefly?" Solicit summary paraphrases or brief restatements to assure sufficient shared understanding of ideas prior to actual evaluation.**

In the heat of the storming, ideas may be abbreviated, put into unfamiliar language, or bent away from the original intention. The clarification time is just that, a time to make option

tion. The clarification time is just that, a time to make option statements clear enough to measure. At this point, the proposals just need clarity, not spit and polish. Mutual understanding is the task.

2. Sort into Themes and Cull, if Necessary

> **Break options into themes or categories if the list of generated ideas is long or overlapping. If appropriate, select the most promising or intriguing for further review.**

Storming sessions often generate many ideas. The adventure of coming up with all those ideas can fade if a group is overwhelmed with sorting through the options to make sense of them. Bug-eyed excitement can turn into bleary-eyed boredom, unless the categorizing and culling process is rewarding, too.

3. Weigh, Choose Option(s)

> **Decide on the method to select one or more options for refinement, a new burst of thoughts, and/or action planning. Measure the ideas against themselves and against solution requirements or wishes, then choose the proposal that will give you the most effective outcome.**

This is decisionmaking time. You can choose from a variety of methods to end up with quality, high-chance-for-success ideas. The techniques range from the quick-and-dirty to the more deliberate and time-consuming.

Don't be fooled into thinking that you only need to apply an elaborate evaluation scheme and you'll end up with a good result. Many ideas are complex. They need refining and a sound understanding of what they entail. People who are going to implement the solution need to have ownership in it. Therefore, the right people have to be involved in the decisionmaking and action planning.

4. Search for More Ideas

> Go back to the search phase—if you realize that the ideas you generated don't fully fit the requirements.

When you generate proposals, you often learn more about the current and the desired state. This new information may cause you to adjust the outcome you need. And a new or refined target could require new directions in the search and a set of brand-new ideas. In addition, you may need to go back to cooking up more proposals after finding that your original outpouring of options was too tame. You may need to try further idea-generation techniques that will give you more creative power.

5. A Second Look: Analyze Potential Problems

> Analyze the forces that will help the option work and those forces that will impede its chances of being successful. Conduct a force field analysis ("Analysis Methods," Chapter 5) of the pressures bearing on the proposal. Determine if the possibilities are greater than the hazards of carrying out the option. Use the results of the analysis to take action in the next phase.

Choosing and getting a good idea on the table is important. To maximize the chances for the chosen proposal to work, though, you need to take a second look at it. Most choices of new ideas have impediments as well as possibilities. Before you take action, lay out the forces that play on the option.

Force field analysis is an excellent tool to use here. You might want to look for forces in the big picture (sponsor, client, technical community, or user acceptance; funding; developmental sources). Look at the realm of the idea itself (patent infringements, technology availability, ease of development or production, or inherent concerns). Include forces in yourself (commitment, other pressures, knowledge, or skill).

EVALUATION METHODS

When you evaluate ideas, you have to be on your mental toes, maintaining both a soft focus and a discriminating eye. You have a number of handy tools to help you study the proposals. Each level of screening has techniques to aid you. Figure 3 shows an assortment of assessment methods for the three levels of screening. Each of the evaluation methods will be explored below. Some don't need much explanation. Others are illustrated for better understanding.

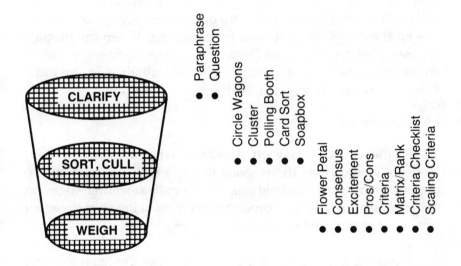

Figure 3. Evaluation methods by screening level.

Clarify

Paraphrase

Briefly restate, in your own words, what you think a stormed idea means. You'll be able to quickly discover how well your understanding matched the intention of the proposal's author. It's much better to suggest an interpretation than assume you know and miss a jewel of an idea.

Question

Ask for a clarification or interpretation of an idea. A lot of discussion is not appropriate at this time, for you want to keep the mood open to all the ideas, not home in on a few. You are pushing for understanding. Thus it would be tacky to say, "What on earth is that pea-headed idea supposed to suggest?"

Sort and Cull

You could sort and cull your ideas at the session in which they were generated, or undertake the evaluation and testing at a later date. Sometimes allowing the mind to incubate overnight will cause more ideas to bubble up or will give you good perspective on the categories you need for clustering. When the proposals are generated by a team, however, group members are often anxious to at least clarify, categorize, and cull ideas in the same session in which they generated them.

Circle the Wagons

At the end of a day on America's wild, western frontier, pioneers would circle the wagons to focus their energies and to protect themselves. Individuals or teams pioneering ideas in storming sessions have to protect ideas from being shot down by errant judgments and discussion.

If you are a facilitator or group leader, take the team through the list of newly formed ideas. Ask the group for a quick assessment of whether each idea is workable or possible. Even if only one person thinks there is possibility in an idea, you circle that item. This allows people to see the merit in the stormed ideas and also focuses the energy of the group.

If there is a fair amount of divergent thinking, you could have several individuals make an evaluative pass through the listed items. Have each evaluator use a green felt-tipped marker to circle those ideas they think could work and a red marker to circle options they believe are too impractical or don't have a ghost of a chance of working. Space out the evaluators so they aren't working on top of each other.

Those items that have both green and red circles could be discussed at that time or added to an "issues" list for later dialogue and resolution.

Cluster

Long lists of ideas may need to be grouped, combined, or categorized to make them more manageable. You'll have to do this without losing the brilliance of specific suggestions.

Put ideas into clusters. You may want to make just three groups: gems needing refinement, too off-the-wall for now, and sound possibilities. As you survey the items, you might find overlaps where ideas could be combined. Your ideas could be lumped together according to the specific target they solve. Or themes might develop naturally. Table 1 shows one way ideas from the case of the flourishing environment could be clustered.

One of the most potent sorting devices is the mindmap. This technique was described in Chapter 7 as a storming strategy. It can also be used as an arational, nonlinear structuring technique. Use the mindmap format for quickly arranging a host of ideas around categories or themes.

You can use the mindmap as a way to not only sort ideas but also to take them to a new dimension. Extract categories out of the listed ideas by transforming them into another framework—into mechanisms, methods, processes, or assumptions. Put the word methods, *for example, into the nucleus of the mindmap, then build branches and spines out of the methods you see in the original list of ideas. This allows you to get to the skeleton on which the proposals hang.*

By developing the categories and themes, you can build shared team understanding about the proposals. This process can also give you new information about the current conditions and the desired state you were exploring in the target phase. With new data you might want to revisit the target and search phases.

Table 1. Clusters of Stormed Ideas from "Case of the Flourishing Environment" Used in "Imaging the Future" Strategy Format

Organizational Mechanisms

Rotating creativity council
Slush fund for small projects
Creativity computer network
Budget for creative thinking sessions and backing innovative ideas
Hire guru
Swapping ideas with other departments
Seek ideas from customers/vendors
Pay for one restaurant meal per month to talk about new ideas with co-workers and management
Extravagance/costume day
Isolation periods with no phones/interruptions
Use outside stimuli, personalities, etc.

Physical Environment

Quiet places for incubation
Fish tanks with guppies
Improved employee lounge
Romper room

Management Leadership

Encourage exceeding of responsibility
Executives get, then give, creativity courses
Corporate culture reset
Corporate creed regarding creativity etched on plaque
Freedom to "screw up"
Meeting in the spa
Off-site creativity sessions to handle complex issues
Eliminate executive dining room
Seek ideas from customers and vendors
Invite another company to cross-fertilization sessions

Individual Activities

Check out Walk-mans with stimulating music
Introduce meditation techniques
Recommend healthy diet for good thinking
Calisthenics every morning
Seances
Encourage laughing at self

Polling Booth

Voting can be hazardous to the health of a team that doesn't honor dialogue, dissent, and differences among group members. If your team is operating in the spirit of discussion and consensus, however, polling the members on a first cut through a newly generated mass of ideas can provide some clarification about the shape of thinking in the group.

Have members review the list, then privately vote for an agreed-upon proportion of the ideas. Vote for 10%, 25%, or 33%. I prefer the 33%, for it includes more good ideas and people begin to see the themes emerging. This first step begins with team members privately voting on a sheet of paper. When all have finished their personal list, they simultaneously put marks (with a colored marker or adhesive dot) on items that received their votes.

Voting privately, a key aspect of the process, stops the more dominant individuals from swaying the poll.

While winnowing the original list gives the team a smaller range of ideas to work with, even the reduced list might need a second culling. When three to six items don't clearly stand out, use consensus and combining of ideas. This narrowing of items gives you a less cumbersome number of proposals to weigh in selecting a top choice.

Card (or Stickee) Sort

As ideas are generated, put them on 3-inch-by-5-inch cards, the backs of old computer cards, strips of paper, or large stickees.

When ready for the culling step in "check," simply move the ideas around on a wall, a floor, a cork board. Have masking tape or pins handy. Arrange the ideas into clusters. Discuss the interconnection of the options.

If you didn't want to spend the time on the sorting right after the ideas were developed, you could have the rearranging done between sessions by a group member or a secretary.

Soapbox

After the initial sorting and culling, allow team participants a chance to advocate ideas that have been cut from this round that they believe merit further attention.

Set the stage for this step by saying, "We have been pushing energetically through the creative cycle, including sifting through the pile of ideas we just generated. Since we don't want to screen out proposals that our team should be considering, let's scan the original list. Then we'll create a little soapbox time for group members to suggest or advocate items we could've missed during our first pass through the ideas."

One of the problems of the sorting and culling step in the evaluation process is the possibility of losing the flavor of the original thought. As ideas get mashed together and recombined, the intent or essence might be squeezed out, leaving a more general, homogenized, and bland statement. The soapbox is one vehicle to reinject the taste of the first idea and possibly save a winner from being tossed in the garbage can.

Weigh

If it doesn't really matter what the decision is, add some humor. If you don't want to flip a coin, pick the petals off a flower in the sequence of the number of options you have. The number of the last petal off provides your winner, of course. A daisy works better than a dandelion! But you most likely want to be more exacting, which you get in the methods that follow.

Consensus

At every leg of the trip through the creative cycle, openness and effective interchange must be the norm. Given a leader who can elicit differences of opinion and guide the discussion, a group working hard to achieve consensus will build solid support for any decision the group makes and will tackle the chosen option(s) with fervor.

Consensus is not a committee trying to design a horse and ending up with a camel. It does entail a group at work, trying to arrive at a decision with which all members of the team can live. It isn't a snap. It takes time. It requires team members who lay data and feelings on the table, push for shared understanding, and believe in win-win decisions. Here are some thoughts to guide the consensus-building process:

- **Focus on issues,** not personalities. Seek the logic of the discussion. Understand the emotions which drive the dialogue. Share, don't argue.
- **Avoid conflict-smoothing techniques.** Don't squelch honest differences of opinion, for unresolved issues will often come back to bite later. Pouring oil on conflict may keep

valuable information hidden from the team, as well as set up polarized subgroups.
- **Listen** carefully to the views of others.
- **Avoid voting** where the majority get to rule. Voting, when used to control discussion, inhibits the development of group cohesion, a vital ingredient of a highly productive team. It sets up winners and losers on the team.
- **Encourage others** to air their views, especially those who tend to be silent.

Ideas forged through consensus move more quickly through a system. When people believe they have had a hand in the development of a proposal, they will put more zest and muscle into helping it survive.

Excitement and Gut

Sometimes this method can be as simple as choosing an option based on its emotional appeal. When you desire high ownership and participation in planning and carrying out the creative thought, you want an idea that will excite and motivate—you and others.

In some of the search formats, you were asked to choose an idea for how it felt, not how rational it appeared.

In Chapter 7, participants in the analogy storm illustration were asked to come up with an intriguing or exciting metaphor. "A spider spinning a web" was the analogy they chose. Choosing an option that was dull and boring would not have led the group to become fully involved with the analogy. In imaging the future, members in the storming process were asked to choose images that excited them about how creativity was being fostered back home.

Just as some of the storming strategies ask for intuitive, feeling-oriented choices, the check phase also allows such responses. You could, for example, query team members about what option most filled them with intrigue, excitement, adventure, or fascination.

Pros and Cons

One of the oldest methods of evaluating ideas is to simply develop two lists for each idea. One list summarizes the advantages, or pros, the other presents the disadvantages, or cons. While one of the proposals may clearly stand out as containing all the characteristics to be a winner, many creative thoughts have pluses and minuses. Often a simple listing smears the differences between the options because you can't tell the quality or impact of each pro or con. Weighing each advantage and disadvantage strengthens the process.

Criteria

During the target phase, I cautioned against too quickly settling on special requirements that new ideas would have to meet. Minimal criteria might have had to be proposed then to deal with unique constraints. By building too firm a measuring stick at that time in the creative thinking process, however, imagination could have been put into a tight-laced corset, restricting the inventive, natural flowing of ideas.

Now is the time to reexamine any criteria developed then, to see if they are still standards against which to judge the ideas. New criteria can also be suggested at this time to help show what freedom and what constraints lie along the path to finding solutions for the desired state.

Criteria strengthen the selecting process by setting a template against which to judge each option. They *are characteristics, standards, identifying marks that can be used to gauge whether a new proposal has the needed traits or qualities.* A criterion may be general, like "marketable," or specific, such as "would fit the needs of 35-to 55-year-old consumers." Evaluation of some criteria requires discussion and consensus. Others mean gathering more information. Samples of criteria are:

- **Meets key objective:** Moves us from 15 to 30% market share; protects the environment; builds on or expands existing company products; keeps departmental projects in $2–5 million range; meets at least four of five governing principles

- **Feasible:** Can be completed with existing equipment and minimal capital investment; uses proven technology; techniques are familiar to operators
- **Growth-oriented:** Will fill a need not currently met by competitors; can get through the blood-brain barrier; process/product is state-of-the-art
- **Sound:** Congruent with XYZ theory; builds on existing products; uses present production equipment; can be accomplished in available time frame
- **Solution-oriented:** Will solve problem for next six months; removes the primary cause/deficit; reduces side effects; will allow us to stay in the critical path

In the case of the challenged O-rings, several criteria could have been developed to determine the most promising ideas generated in the brainwriting session. Here are a few possible criteria:

- *Enhances sealing of rocket joints*
- *Improves safety*
- *Minimizes delays to schedule for planned space flights*

Sound judgment is needed in applying these criteria. Both technical and project management considerations are important. How much does the design need to be fixed? How long will it take to investigate the full merits of a suggestion?

Development of esoteric materials and designs were not necessary to fix the problem. A third O-ring and interlocking insulation were the ideas actually used to get the space program back on track. Longer term solutions to provide versatility for space flights under a wider range of weather and stress conditions could, of course, also be chosen to be worked on away from the critical path.

Criteria make the option choosing more exacting. They allow for ideas to be checked individually to see if the options match the standards, requirements, or targets. Simply developing a checklist for how well ideas fit each criterion will generally make the option-choosing more rigorous.

The evaluation process can become more effective when an

additional step or two is instigated. One step lays out the data for evaluation on a matrix. A second weighs each option criterion. Often the steps work together, for a good display of options is helpful in assigning weights. The two key evaluation steps are shown graphically in Figure 4.

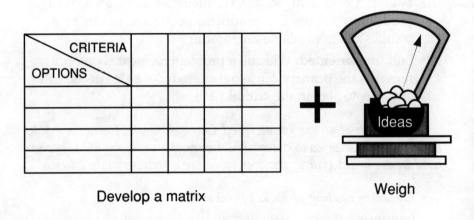

Figure 4. Increase effectiveness with a matrix and weighing of ideas.

A matrix exhibits the ideas and criteria so weights can be applied to the options. By putting the options in rows and criteria in columns of a matrix, for instance, you can contrast the ideas. We'll explore three ways you can use a matrix to weigh ideas:

1. Vote for a small portion of the ideas and rank your favorites.
2. Mark the criteria that match each option.
3. Use a scale to measure all ideas against each criterion.

The method you choose will depend on the degree of rigor you need to arrive at a sound evaluation. There are positives and negatives with each one.

Matrix with Ranked Votes

In this first technique, you create a matrix. Then you vote for your top choices, placing them in a ranked order. Although the

CHECK: ASSESSING OPTIONS

least rigorous of the three approaches presented in this section, it may be appropriate when you don't need to apply specific criteria.

Compose a matrix with four columns. Since this method requires a group, develop the matrix on newsprint that has been posted on the wall for everyone to see.

Put five options, or another manageable number, in the left hand column. Three of the columns will contain a weighted rank. Ranking two to three ideas out of five should provide an initial, collective view of how your group evaluates the different possibilities.

Have everyone privately choose their top three choices, then put the choices in a ranked order. The top choice gets a weight of three, the second choice a weight of two, the third choice one. When everybody in your group has completed choosing and ranking, all team members go to the newsprint at the same time, putting a dot or check mark in the cell that will give the appropriate weight to the choices they made.

Since each dot in each cell receives the numerical weight for that column, multiply the dots in each cell by the weight ranking of the column, then total by rows. The totals will give the group weights for each option.

Sample: Imagine being included in the group of individuals assigned to generate fresh, innovative ideas to solve mythical DesertSun's thirst for water. You generated ideas using the picture tour storming strategy. Then you and your cohorts culled your original idea list to five. Figure 5 shows how six people on your team might have ranked the five options. Your group chose the home purification system as their top choice. But a government-sponsored conservation program with major inducements was not far behind.

Discussion of the ratings would help show why people made the choices they did. Based on the dialogue, the team would have a clearer picture of what the assessment meant. The ratings and choice of the group could change as a result of consensus discussion.

RANK / OPTIONS	First (Weight of 3)	Second (Weight of 2)	Third (Weight of 1)	Total
Underground houses			● ●	2
Home purifier system	● ● ●	● ●		13
Anaerobic toilets		●	● ●	4
Ice from Alaska	●	●	● ●	7
Induced conservation	● ●	● ●		10

Figure 5. Example of a matrix with ranked votes.

Criteria Checklist

A second matrix/weight technique simply checks to determine if each idea meets a specific criterion. The strength of this technique is in the use of criteria. Its weakness lies in not determining the degree to which an idea fits each criterion. This level of evaluation might be quite acceptable for what you need.

Develop criteria. Lay them out on a matrix. Review each option to determine if it meets the quality or requirement of each criterion. You make one of three responses. A check means the option fits the criterion. Leaving the box blank means there isn't a match. A question mark suggests you aren't sure or need further information.

Sample: Contrast this method with the matrix using ranked votes. Again use the five top options chosen by your team evaluating ideas to solve water supply problems in DesertSun. This time, as shown in Figure 6, five criteria were developed. Using this method, your team wanted a collective appraisal of whether your top options matched the criteria. They wanted a judgment on whether each idea was fea-

OPTIONS \ CRITERIA	Feasible	Original	Technology available	Long term solution	Marketable
Underground houses	✓		✓	✓	?
Home purifier system	✓	✓	?	✓	✓
Anaerobic toilets	✓	✓	✓	?	✓
Ice from Alaska	✓	✓	✓		?
Induced conservation	✓		✓	✓	?

Figure 6. Example of a matrix with criteria and yes-no-? values.

sible, original, or had the technology available to develop it. Your team also wanted to see if the proposals could be considered a long-term fix of the problem and had the ability to be marketed, either to the populace or to those in the state who must consider economic and political ramifications.

Once you have completed the matrix, you still have the task of making an evaluation of what the checks mean. One of the problems is that the criteria are all given the same weight. You don't get an indication that not meeting a criterion will rule the option out. However, you have identified areas that require a further look before you can proceed. Or you can let the question marks be negative considerations, areas posing too much work for the time and resources available.

Scaling Criteria

A third approach goes a step beyond the use of a matrix and criteria.

Add a scale that can measure how well each idea meets a criterion. If you want to keep it simple, use easy scales—high, moderate, *or low;* +1 *for a positive quality,* 0 *for neutral, and* -1 *for a negative quality;* 1–5, 0–10, 1–10, *or* 1–100.

Sample: Figure 7 illustrates how the DesertSun evaluation would develop when using a high, moderate, or low scale. The water purifier-recycler and exporting ice from Alaska have the same rating, three *high*s and two *moderate*s. Notice how the weighing picture is modified when criteria are scaled.

CRITERIA / OPTIONS	Feasible	Original	Technology available	Long term solution	Marketable
Underground houses	H	L	H	H	M
Home purifier system	H	M	M	H	H
Anaerobic toilets	M	M	H	H	H
Ice from Alaska	M	H	H	M	H
Induced conservation	H	L	M	M	M

Figure 7. Example of matrix with criteria and H, M, L scale.

Evaluation is a basically subjective process. Hopefully you aren't fooled by the attempt to quantify and to discriminate between options. Although scaling criteria adds more objectivity to a subjective process, dialogue and further refinement of the weights and scales may be necessary to get a satisfactory result.

There is a flaw in the example in Figure 7. Each criterion seems to be given an equal footing. But criteria can have different impacts and values.

WEIGHING. You add more rigor to the evaluation if you weigh the criteria, then weigh the options. Here are a couple of steps to follow to compare the criteria:

1. Divide the criteria into "have to have" or "ought or nice to have" groups. The "have to have" category is for criteria you insist on requiring. You will throw out an option if it doesn't have a specific quality or element. Generally, if the proposal must meet a requirement, you can quantify the desire and have a cutoff amount for keeping or tossing the option. You either go or you don't go with it. If it's a no-go, you toss it and don't consider it anymore. The "ought or nice to have" criteria vary in degree of desirability.

2. Measure "ought or nice to have" criteria against each other on a scale. Use, for instance, 1–10, with 1 low, 10 high. Determine which is the most important criterion. Give it a 10. Assess the relative distance on the scale the other criteria are from the 10. See Figure 8 for an example. It is possible for more than one criterion to have the same weight, including a few situations where there are two 10s.

CRITERIA / OPTIONS	Feasible Meets	Original Wt:2	Technology available Wt:8	Long term solution Wt:10	Marketable Wt:7	Total: Sum of each Score X Weight in row	In each option-criterion cell: Score of how option fits criterion 0-10 / Score X Weight
Underground houses							
Home purifier system							
Anaerobic toilets							
Ice from Alaska							
Induced conservation							

Figure 8. Example of matrix with weighted criteria.

Sample: If your team were to go for a more demanding evaluation than previous methods, you would put your criteria into categories. One criterion stood out as a requirement, as a "have to have." Each proposal had to be feasible. Your group decided to define *feasible* as "judged to be at least theoretically possible by technical members of the task force." All of the final five ideas were judged to be feasible.

The other criteria were in the category "ought or nice to

have." The final five ideas were chosen by your task force because they excited the group, yet you knew the options would all take development to test their full acceptability. You applied weights, shown in Figure 8, to compare the criteria.

Since the governor was pushing for options that would provide a satisfactory solution over the long term, your team gave the highest weight, a 10, to the long-term solution criterion. Originality of options was intriguing and gave you a sense for what ideas were really fresh. As a criterion, however, it was given a lower weight, for decisions would probably not ride on whether an idea was novel.

SCORING. The second part of this evaluation approach that adds more precision is the scoring of the options. Follow these guidelines:

- Evaluate all options against the first criterion, using a 0–10 scale. Measure the degree to which each option meets the criterion, from not meeting it at all, a 0, to matching it fully, a 10. Using the format in Figure 8, place that value above the diagonal line in the cell intersecting the appropriate option and criterion.
- In like manner, continue to score the options, one criterion at a time.
- Once all options have been scored against the criteria, multiply the score by the weight for that criterion. Enter the result below the diagonal line in each appropriate option-criterion cell.
- When all the cells are completed, add the Score × Weight numbers in each row. Enter each answer in the total column to the right of the matrix.

Sample: Note, in Figure 9, how the task force scored the DesertSun options. You have greater definition of the difference between proposals in this evaluation scheme. In using a high, moderate, and low scale for checking the fit between ideas and criteria, three options were tied. In this more precise discrimination among criteria and ideas, the home purifier system became the team's top choice.

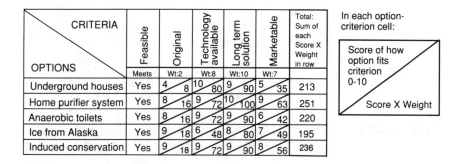

Figure 9. Example of matrix with weighted criteria and scored options.

While this last method gives you a more powerful evaluation, retain the caution that this is still a subjective process. More information may be needed to gain further confidence in the weights. The criteria themselves could distort the picture.

Discussion or review of the process with outside experts or third parties may help you identify issues. Dialogue with the group may uncover hidden agendas or withholding of pertinent data. Such review also helps you know the energy, support, and attention that is available for carrying out the chosen option.

PAUSE FOR PERSPECTIVE

We have just surveyed a range of techniques to gauge how well stormed ideas meet our needs. The methods start with attempts to get clarification, then assist in making the list more manageable. Finally, the processes help weigh the merits of each option. A summary of the techniques is presented below:

Paraphrase Describe idea in your own words to see if you caught intended meaning.

Question Ask author to briefly describe the proposal's intention.

Circle the Wagons Circle items that appear workable.

Cluster	Group ideas to make them more manageable.
Polling Booth	Vote on a percentage of items, to winnow list to three to five.
Card Sort	Collect ideas on cards during storming, then cluster on wall or floor.
Soapbox	Advocate for potential culls before they are thrown away.
Consensus	Push for decision with which all group members can live.
Excitement, Gut	Choose items for their emotional appeal.
Pros and Cons	Pose advantages against disadvantages.
Criteria	Weigh against requirements.
Matrix with Ranked Votes	Use matrix with options in left column, right columns hold rank and vote for favorite options.
Criteria Checklist	Use matrix with options in left column. In right column, check criteria that fit each option.
Scaling Criteria	Use matrix with options in left column. Measure each idea against a criterion, based on a scale.

There are other evaluation methods, of course. The techniques presented in this chapter are intended to be practical, not statistical or exhaustive. Keeney and Raiffa, in their *Decisions with*

Multiple Objectives: Preferences and Value Tradeoffs, give you statistical methods for assessing options that have a variety of targets. Robert Bailey devotes a chapter in *Disciplined Creativity for Engineers* to a formula for estimating the technical potential in new ideas, to see if a brainchild is worth pursuing. For additional decisionmaking schemes, turn to Plunkett and Hale's *The Proactive Manager* and Kepner and Tregoe's *The New Rational Manager*.

You may have a favorite technique that works well for you. The critical point is that you get the level of evaluation and selection of ideas that best suits your needs—whether evaluating proposals for stations in space, sensing devices for steering oil tankers around dangerous shoals, food made out of wood fiber, problems on the shop floor, a new name for a product, or potential uses for a new process. The type of process you choose when assessing new ideas will often depend on the magnitude of the costs, consequences for any mistakes, and impacts for undertaking the option.

When trying to scope out an area, you may need to just be in the ballpark, to just become aware of the direction that will get you closer to your target. Discussion and ranked votes may be sufficient. If you need greater clarity on the quality and usefulness of each option, your decisionmaking should be more rigorous. Weighing criteria and scoring options provides more information and, thus, can lead to a more enlightened decision. Even then you need to make sure criteria are real, understood, workable, and present the factors that actually need to be considered.

Many people stumble once they have made what they consider to be their top choice. Even the best ideas have built-in problems. Implementation can be a swamp, where hidden quicksand can trap you. It's important to take one more look at the option you have chosen to see where you might get snagged or bitten, as well as where you can capitalize on the strengths of the proposal.

Your DesertSun task force chose as their top choice the water purifier-recycler for the home. Conduct a force field analysis to see what will help or hinder you in the implementation quagmire. An abbreviated study of forces is shown in Figure 10.

Re: Developing and using a home water purifier

	HELPING	HINDERING	
10	Potential water saving	Existing home configurations	7
8	Combining of known technology	Homeowner mistrust	6
		Bugs in system	8
9	Support of majority of DesertSun Water Conservation task force	Timeline for developing low cost system	5
		Lukewarm support of subcommittee chairperson	4
10	Long term solution		
7	Foundation grants available	Near term shortage of state development funds	10

Figure 10. Force field analysis of implementation factors.

Since evaluation and decisionmaking is a subjective process, the weights applied to the forces may vary with the judgments and circumstances of the individuals or teams that assign them.

Once you have this analysis, you are better equipped to develop actions, the task for the final phase of the creative thinking process. Try not to get bogged down in this check phase, to lose the uplift and motivation you got while generating ideas. Keep this process moving. Get a decision made so you can take your key thought, your sketch of an idea, your outline and get on with it.

COMMUNICATION TIPS POSTSCRIPT

Caution and care are needed when checking out the worth of ideas. Since one of the core rules of storming is to allow wildness and getting all proposals out, some of the ideas that are proposed can seem to be too far out or too fragmented for full consideration by the group. An attitude that is heavily critical at this point could be detrimental, for good but undeveloped ideas or even strange options may have real merit if a little time is taken for greater understanding. Also, overly critical critiquing can cause people to shrivel their creative thinking for the next idea-generation round.

Using good communication techniques during the checking process can greatly enhance its value. Here are a few key ones:

- **Reaffirm what will happen to the ideas.** Remind the group why the proposals were generated, how they will be used, and who owns the ideas—a sponsor, client, team, problem solver, or specific individual. Idea generators often feel their suggestions are a part of themselves and retain a sense of ownership in their proposals. Allow their thoughts and role to be honored.
- **Clarify ideas with a paraphrase.** Test your assumptions about options that have been presented. Show that you honor an option by putting your understanding of an idea into your own words and sharing the paraphrase with the group or generator of the proposal. This restatement will help assure that you get the essence of someone's thought, for you will receive a nod of agreement or a clarification. Many ideas get thrown into the wastebasket because they have not been understood. Also, people sometimes need to hear how others have perceived their ideas to get further insights into their own thinking and ability to convey thoughts.
- **Point out at least two positives before you make a negative statement about an idea.** This will help restrain hair-trigger judgments and will show that you have tried to put your comments in a broader framework. But you have to be genuine in your positives. Nonverbals such as frowns, raised eyebrows, or caustic tones belie true intentions and thoughts. Masking negative reactions with mouthed positives, while wanting to say that the idea is the stupidest thought you have ever heard, will come out negative, at least in body language.
- **Show how idea fragments or seeming losers can be turned into winners.** Look for the core or golden nugget in each brainchild. Show how the essence or beauty of someone's proposal, even though somewhat hidden, can be brought out, reframed, or combined with other ideas.

- **Draw the idea, when appropriate.** You can do this during the clarifying period, when the team is pushing for a shared understanding. You will give people an actual as well as a mental picture of what an idea was intended to be.

These communication tips could easily have been presented in the target phase. Analysis of the situation requires a good flow of information, as well as an evaluation and categorizing of data. Paraphrasing, pointing out positives, and helping ideas become winners are especially pertinent to the check phase, however.

Individuals who have ventured new ideas—especially those which might seem to be crazy, nonsensical, or wild—often feel vulnerable. Most people don't want to have their peers consider them strange or out of step. Checking ideas requires evaluation and selection. Yet it must be done in a context of honoring all proposals generated during the search phase.

9   Action: Launching Ideas

When you return to your parked car on an icy day and the lock is frozen, the vertical thinker may try to heat the lock with his lighter in the wind. The lateral thinker may shelter and heat his key with the flame.   —EDWARD DE BONO

Creativity is not pretty. You have to go through a lot of failure before you come to your success.
—PAUL MACCABEE

Once a promising solution idea is chosen, there is often still much work to be done.

In 1983 the now famous Nobel laureate K. Alex Müller, an IBM researcher working in Switzerland, attended a conference in Erice, Sicily. He was stimulated by a lecturer who noted that the molecular structure of ceramics was similar to that of existing superconductors made of metal. Therefore, conjectured the speaker, ceramics might also be a superconductor candidate.

While walking in a monastery garden, Müller was inspired to pursue ceramics as a superconductor. Since he didn't work in ceramics, the researcher didn't have the shackles of conventional thought.

Müller's lab partner, West German J. Georg Bednorz, joined him in the pursuit. Testing oxides to speed the flow of electricity seemed far-fetched, for such compounds were commonly used to slow down electrons. To pursue such an off-the-wall idea, the pair had to bootleg the time to research the oxides.

The two worked hundreds of hours experimenting with oxides, sometimes into the wee hours of the night. Although doubts arose from time to time, they continued their search. The turning point came when Bednorz came across a new oxide in a chemistry journal. He tested it and eureka! It worked. Each time Bednorz and Müller tested it, they got the same positive results.[1,2]

One of the remarkable aspects of their story is the work they did before the idea became a reality. Considerable effort was needed to get from that inspiration in the monastery garden to the replication of their results at the University of Tokyo within days of the publication of their findings. The road to uncovering the right oxide and conditions for high-speed conductivity was paved with persistent belief and much activity.

When an idea comes in a flash or through the synergistic effort of a team on a retreat, it may be well cooked and ready to use. Or the proposal may be partially done. The idea could come as a direction in which to look, a skeleton needing flesh, or an uncut gem requiring skill and a practiced eye to bring out its beautiful facets without destroying it. Müller's imaginative flash was a direction in which to look. Painstaking setting up of experiments, choosing oxides, and determining the conditions in which to test the compounds were parts of the laureates' action phase.

STEPS TO MAKE THE IDEA REAL

The *action phase* is a step in the creative cycle that *shapes, plans, tests, and implements the chosen option.* It may include selling the idea to others. People who will be affected by the solution, or their representatives, need to be brought on board early. You need them to provide input on the situation and criteria, to help generate and evaluate ideas, and to help develop steps to get the proposal moving. This action process must assure that the right people have been or will be involved.

The action phase doesn't have the range of fancy analysis, idea generation, or evaluation and decisionmaking techniques found in the other three creative cycle phases. There are, however, steps, cautions, and guidelines to be considered in completing this phase. And the force field analysis, begun in the check phase, needs to be completed in this final process. A sketch of the phase is portrayed in Figure 1.

ACTION

- Develop plans to enhance option(s) and reduce potential problems

- Implement the proposals, tests, models, or simulations

Figure 1. The action process.

In this process, you plan the actions or series of activities that will launch the option(s) chosen in the check phase. And you set up acts that will protect it over time.

1. Develop Plans to Enhance Option(s) and Reduce Potential Problems

>Based on the force field analysis completed in the check phase, preview the forces. (1) Determine the main activity or set of actions that will help the chosen option(s) become real. (2) Plan steps that will protect and nurture the action proposal(s), capitalizing on the helping forces and finding ways to minimize the hindering ones. Steps 1 and 2 may be done concurrently.

Nail Down Critical Tasks

From the force field analysis, you have a jumble of factors to wrestle. Some forces shove you toward your goal. Others tug on your sleeve, clamoring for attention, like a pest you wish would just go away. You have to make sense out of the forces, finding a way to reduce or eliminate the power of some factors and to build on the strength of others. You will have to determine the critical one to three things that will assure the successful implementation of your chosen idea. You want to tie down the main thrust of what should be accomplished. There are numerous examples of what you might need to do.

Develop the assumptions or hypotheses under which you will proceed. Perform the appropriate calculations. Pull a team together to map out the project in detail. Set up an experiment. Build a model—a sketch, a computer modeling program, a mock-up. Probe the literature. Co-opt a sponsor. Talk to colleagues. See an expert. Sit down with a lawyer to establish the idea trail for patent registration. Walk to the lab or assembly line or design room to test the idea. Create a market plan for selling your idea inside or outside the organization.

With any action planning, several items should be included in each proposed activity:

- Tasks to be completed and by whom.
- Standards to follow. Methods to use. Resources to be tapped.
- Processes for monitoring and evaluation.
- Timeline with milestones and, if important, inchstones.

In the case of the parched land, the mythical southwestern state of DesertSun determined to find fresh ideas to help solve, on a long-term basis, some aspect of their scarcity of water. The group assigned to generating proposals was intrigued with several ideas. They chose the home water recycling and purification option as their number one place to begin focusing attention and funds. Their main action was the following:

Develop a practical, moderately priced home purification system for recycling appropriate types of waste water. Assign the technical leadership for the recycling system to DesertSun University. Have the State Water Division set standards, participate in the identification of assumptions under which the system would operate, and monitor its development. Prepare a project plan, in conjunction with the university, detailing the projected scope, schedule, and funds needed.

Sometimes your action steps are clear and can be quickly listed. At other times, when you aren't quite sure what would work or you would like additional thoughts, you need another burst of ideas.

You can go back to search methods. Keep the approach and mood an imaginative storming one and you'll collect a good range of action proposals. Brainstorming, for example, often works well at this point, for you generally just want good, not necessarily novel, ideas. Choose the key ones. Keep the actions bite-sized.

Back Up the Main Plan

Once you have nailed down the critical tasks, you are ready for the next step.

Create additional actions that can build on the helpful forces and reduce inhibiting ones. The forces identified in the force field analysis vary in their power to assist or resist the idea as it exists in the current state. Develop contingency plans. Set up a program for handling problems inherent in the option and resistance to the proposed idea or change.

Here is a method for backing up your main plan by tackling the field of forces and by pinpointing actions:

1. *Determine the changeability of each force.* To move the chosen option to the desired or target state, the forces have to be altered. Some of the factors, however, may not budge, may not be capable of changing. It would be futile to try to push

a force like the Rock of Gibraltar off its base. Climbing over it, tunneling through it, ignoring it, or neutralizing it might get you further. Look for forces that have the potential of being changed, regardless of the impact they exert on the current conditions.

2. *Storm actions to reduce negative and enhance positive forces.* To back up and protect the main action plan, review the force field analysis. As you look at the display of forces, generate good ideas to strengthen or add to positive factors, as well as options to eliminate or lessen the force of inhibiting factors. Remember, according to Kurt Lewin's theory, you can't just stack up positive forces. To get movement toward your goal, you also have to reduce the hindrances.

In the case of the parched land, the primary actions for launching a water recycling and purification system were described. A few possible backup action proposals might be:

- *Have DesertSun water conservation task force members develop a network of political and financial support.*
- *Contact foundations to learn of their requirements for submitting proposals.*
- *Create budgetary phases, where the proportion of foundation and government funds change in each phase.*
- *Solicit home real estate developers willing to use the system on a trial basis.*
- *Involve homeowners in the development and refinement of the system.*

There are countless scientific and engineering ideas proposed every day. Many of them are much more complex than the simple illustration I'm using here. The nub of an idea may come from an individual or from a host of sources. Each proposal requires action by one person or many.

A neuroscientist who comes up with a bright idea for setting up an experiment to test for a hypothesized neurotransmitter might be the only one involved in the action plan. The search for a new atomic particle could also call for a delicate experiment but could require a team to handle all the planned tasks. When ideas get turned into megaprojects, like a proposed superconducting super collider or space station, the tasks and technical work just to reach a suitable design can escalate considerably.

Once an idea gets to the magnitude of a project and action steps start colliding with each other, charts that lay out interrelating activities and a critical path are needed. Since project management is a field of its own and there are many books and courses on the subject, that level of planning goes beyond the scope of this book.

2. Implement the Proposals, Tests, Models, or Simulations

Activate the plan. Monitor how closely the activities fit expectations and the critical path. Make midcourse corrections. Hold a postmortem after the execution of the idea.

This step is for rolling up sleeves and getting on with the work. It's doing what we said we were going to do. Since situations and events keep changing, we have to track how closely we are to what we expected to happen. We may have to alter our plans. We have to be alert to new possibilities, even as we are carrying out our original intentions.

A soft focus even while we are paying attention to our original plan could lead to serendipity. That is what happened at the pharmaceutical company Upjohn. Researchers were on the trail of a high blood pressure drug called minoxidil. Although their tests showed the drug worked, it had unwanted side effects. The oddest effect was the growth of hair.

Minoxidil was about to be thrown away when someone focused on hair rather than blood pressure. By redirecting their attention, the researchers learned more about the chemistry that makes hair grow. With this redirected effort, Upjohn now seems to be on the way to developing a hair-growing formula that could make bald men eternally grateful.[3]

CONVICTION AND CAUTION

While there are high stakes in reaching toward new products, scientific discoveries, revolutionary processes, and novel designs, there are also very high risks. There can be heavy costs in time, shrinking funds, thwarted careers, and dead ends. Our chances for successfully introducing an idea into an organization or a technical community *can* be increased. We need two attitudes, seemingly at odds but actually complementary. Both conviction and caution are needed.

In the first attitude, conviction, we reach out enthusiastically and with certainty in a new direction. Inventors often cling to the belief they are on the right track. Their vision keeps them at painstaking tasks. It is boundary-pushing boldness that carries them over rough spots.

When Oliver Evans was told he would be declared insane if he continued to talk about developing a steam-driven carriage, he persisted and invented a steam-powered amphibious dredge.[4]

K. Alex Müller and J. Georg Bednorz worked the fine line between perseverance and disaster—bootlegging experiment after experiment on the belief that oxides might produce the mother lode for high-speed conduction of electricity.[1,2]

A self-taught machinist, Stanford Ovshinsky, proposed the use of noncrystalline materials for semiconductors. He was booed by participants at a technical meeting when he presented the idea. Other scientists were intrigued, however. Ovshinsky continued on his pursuit of technical questions, earning over 100 patents. Today he has a thriving company which manufactures polycrystalline solar cells.[5,6]

Another attitude, caution, increases our chance of success. Boldness and hesitation may not seem to mix but are parts of the mental flexibility needed to think creatively and wisely. Caution holds us back from blindly plunging ahead. It keeps us wise to how our minds can trap our thinking, as well as propel us to new thought. We need to be cautious in our convictions.

There are numerous examples of being locked into existing models of thought.

Phrenology, which assigned mental abilities to very specific areas of the brain, held sway in parts of the technical community in the nineteenth century, even while contrary evidence was being formed. Also, over the years wondrous patented gadgets and medicines have been acclaimed by medical doctors as having amazing healing powers, while their colleagues call the devices and preparations ineffective.

We need to be cautious about our beliefs because we may be wrong. We could be trapped in our own paradigms. Or we may not have all the perspective, data, knowledge, or skills we need. Or we may be caught in the omnipotence of an idea that blinds us to contrary data or the need for help from others.

Caution is needed along with conviction because even breakthrough ideas can be driven by jealousy and obsession. Without a balance of conviction and caution we can cause blindness or distortion toward our role or that of others in the discovery process.

Michael Bliss, a University of Toronto historian, has written **The Discovery of Insulin,** *a book describing the tumultuous route that was taken to find insulin.*

Popular accounts of the day had two unknown scientists ferreting out the secrets of insulin. Frederick Banting, a 30-year-old surgeon, conceived the critical piece of the puzzle while he was trying to fall asleep reading a medical journal in October of 1920. Charles Best was a 22-year-old whose contribution to the quest was doing the chemistry.

Apparently the real story was quite complex, with others entering the picture.

According to Bliss, there were jealous bouts and an apparent withholding of data. Banting knew that the two couldn't carry out the experiments alone. Best didn't want another chemist, James Collip, to join them and share in the race for the prized goal. But Banting persuaded Best to relent. The head of the physiology lab, J. J. R. Macleod, gave the team advice. After stormy threats from Banting that he would take his research to another center, Macleod gave the cantankerous scientists more resources.

When the team was satisfied they had found a way to reduce the blood sugar in animals, they assigned the purification of the formula for human use to Collip. A talented, intuitive chemist, Collip found the right technique in 1922. At first, with the permission of Macleod, he withheld the process from Banting and Best. Eventually and fortunately, the four were able to set aside their differences long enough to finish their work.

The Nobel Prize went to Macleod and Banting. Macleod is said to have commented, "If every discovery entails as much squabbling over priority, et cetera, as this one has, it will put the job of trying to make them out of fashion."[7]

The pursuit of discovery is certainly a very human endeavor, with fame, egos, financial reward, honor of colleagues, and future work riding on the effort. In the spirit of conviction and caution, honor not just your own ideas, but the many seeds of thought and the supporting contributions of others. Credit your contributors and sources as much as possible. Periodically take the pulsebeat of your action plan, your team, your own involvement. Fix weak spots. At the end of an important milestone, stop to celebrate or have a wake. Conduct a postmortem to identify lessons learned for building your next or your ongoing venture.

COMMUNICATION TIPS POSTSCRIPT

When relating to others, the communication guidelines for the previous creative cycle steps can be very useful in this phase, too. If you are doing the action planning in a meeting or similar group session, add the following two tips:

- **Push for manageable steps, clear responsibilities, and specifics.** Many staff meetings and retreats expend much energy coming up with proposals. But those efforts, although generating creative sparks and enthusiasm, often vaporize later because the actions weren't tied down at the originating sessions. Pressures on the job and lack of clarity about what is to be done can cause good ideas to evaporate.
- **Evaluate the meeting's effectiveness periodically during the session or at its conclusion.** Then correct or discuss any

problems in the processes you have been using. Also take time to reinforce techniques that work well.

Often, during a meeting, in sharing your idea with others, or in laying a patent trail, you need to make the concept or idea as *visible* as possible. Machinists, model makers, patent attorneys, or potential sponsors need to see with their eyes what you see in your mind's eye. Making things visual with sketches and drawings was highlighted earlier, especially in the discussion of visual media in Chapter 2 and in the communication tips of Chapter 7.

Countless inventors have created drawings of their inventions to make the uniqueness of their ideas tangible. Harold Rosen, author of a number of patents, including the first workable communication satellite for outer space, was no exception. Patent drawings were a regular part of his action planning. Figure 2 shows the sketches for one of his inventions, a directional antenna system for satellites.[8]

In the action phase, you often have to get your ideas across to others. For many scientists and engineers, selling their proposals is a nuisance. It almost seems unnecessary. They often believe their ideas are clear and obvious, worthy of being implemented without fanfare or having to convince someone.

The psychological types we studied in Chapter 2 give us clues about how people can feel their ideas should be accepted automatically. Many of us, myself included, have a preference for using our intuition and thinking, a mixture that gives us big picture flashes that we think are logical and obvious. We use our imagination to create new designs, processes, theories, and systems. Then we build a model or concept, in our mind or on paper, to logically tie even the most complex creation into a neat package that should be understandable to everyone!

I used to be baffled when others did not immediately see the wisdom of my ideas. When someone wasn't able to see, touch, or feel my thoughts, they weren't dense or dumb. I just hadn't stepped into their mental framework to see how my thoughts sounded to them. So I couldn't choose words or an approach that would make sense to them. When you want to sell an idea to

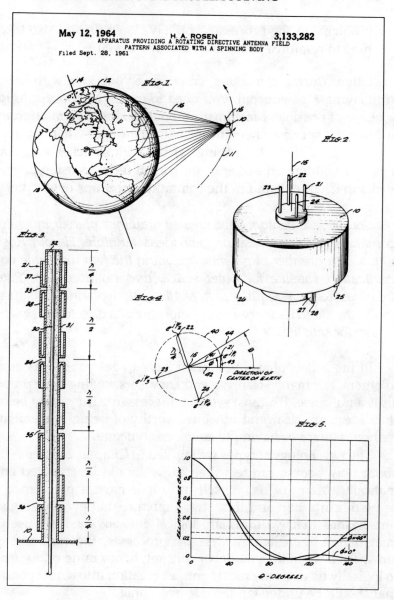

Figure 2. Harold Rosen's patent drawings for a satellite's antenna system. From *Inventors at Work*. Reprinted by permission of Microsoft Press. Copyright 1988 by Microsoft Press. All rights reserved.

someone, you have to tap your understanding of the different styles people use to process information.

Review your own preferences. Speculate on the preferences others might have. Anticipate how they might approach the idea you have to offer. What might they like to hear or know or see to make the idea palatable to them? To sell ideas and their implementation actions, develop empathy for the way other people approach the world. Learn to shape your behavior to be understood by someone else. For clues on selling, review the notion of psychological preferences, described in Chapter 2.

In my creative thinking workshops, I sometimes have participants get into small groups according to how they prefer to relate to the outer world (judgers or perceivers) or where they get their energy (extroverts or introverts). From the point of view of their strong preferences, groups create and give presentations following the modes they would like others to use in selling to them.

Based on data from those high preference groups and from concepts and research on psychological types, here are a few, but important, tips to use when trying to get your ideas across to others.

- Extroverts like enthusiasm, conviction, interchange, and varied media, with proposals clear and up-front.
- Introverts prefer a presentation that is well thought out, not flashy, low pressure, with background or introductory material beforehand or afterward. They want to reach conclusions at their own pace, usually after the session.
- Judgers like organized thought they can see in the format of the presentation; conclusions; snappy data trimmed to the essentials (e.g., items with bullets).
- Perceivers desire a range of data that allow them to draw their own conclusions and a chance for give-and-take. They can often handle data and formats that are open-ended and incomplete.
- Frame your ideas and action plan in a mental function that is the decisionmaker's strong suit—practical results and

facts for sensors, possibilities and the imaginative angle for intuitors, logical sequence for thinkers, and values inherent in the action for feelers.

Knowing the other person's type, especially the key decisionmaker's, isn't mysterious.

Reread the section on type in Chapter 2. Learn what you prefer. Observe those with different preferences. Ask the decisionmaker what he or she would like to have included in a presentation. Ask others who have been successful in presenting ideas to the decisionmaker what style seemed to work.

Make sure, however, the person you are targeting as the decisionmaker actually is. The real decisionmaker could be someone other than the person you think—a valued lieutenant, someone off to the side with political clout, a staff expert, a quiet sponsor who stays out of the limelight, the head of a rival committee. Ask several people how decisions are made to help you learn who has to hear your pitch. You may even want to touch bases with several people prior to a meeting to gain wider support before you actually make the presentation to "officially" sell your idea. Communication, of course, is never ending. There is a new beginning with each fresh idea that emerges from our imagination and needs to be shared in some form.

We now turn away from our specific examination of the creative cycle. Our focus will be on the more general approaches that can enrich imaginative thinking for individuals, groups, and the whole organization.

Seeds of thought, on their natural course, may grow into sturdy seedlings. But we may want more. We might seek the novel, the breakthrough. That often requires us to do more than graft ideas together. To find fresh forms, untried paths, or new combinations, we may need to stimulate, tease, prod, and enhance our minds and our teams.

10 Enhancements: Enriching Individuals and Groups

> *Whenever a knight of the Grail tried to follow a path made by someone else, he went altogether astray. Where there is a way or path, it is someone else's footsteps. Each of us has to find his own way.... Nobody can give you a mythology. You'll find the images in your dreams, your visions, your actions—and you'll find out what they are after you have passed them.* —JOSEPH CAMPBELL

> *An effective thinker can turn thinking skills off and on, and focus them on a wide range of subjects and situations.* —EDWARD DE BONO

A lone figure treks up the mountainside in an exotic land. In search of wisdom and personal illumination, she struggles higher and higher along the rugged peaks. At last she reaches the cave of the Great Guru, the one known far and wide for enlightenment. The guru cultivates depth of insight in his cave retreat. The woman, having survived the tortuous journey, can now probe the mind of this wise one.

This or a similar fictitious scene, often portrayed in cartoons with a zinger that pans either searcher or guru, is symbolic of our search for an extra quality or state of being. We may journey—mentally, spiritually, or physically—far and wide to gain an added measure of wisdom. Often that for which we are looking is within ourselves or close at hand, available without such an external journey.

If, after having taken the trip over this book's terrain thus far, you want an extra imaginative source or catalyst—an afterburner thrust, this is the chapter for you. The foundation of the creative mind and process has been laid. Now is the time to explore how to give an added boost to your own inventive reasoning or to your work group's attempts to be innovative. Thus far we have traveled over the creativity landscape of:

- Elements of the human instrument that are the keys to creative thinking: (1) Balanced rational and arational thought, blending logic with intuition. (2) Effective mental habits and media for expressing our thoughts. (3) A relaxed, flexible, soft focus that can move smoothly from the practical to the outrageous. (4) Childlikeness that maintains curiosity, playfulness, fantasy, and emotional color.
- Creative thinking broken into four phases—target, search, check, and action—for stepping through the imagination-producing process.
- Strategies for igniting creative thought: (1) Blueprints using storms of words, analogies, pictures, and mental images to break mind-sets. (2) Relaxation and mood music, dreams and free association.
- Suggestions for tuning the human instrument and for igniting the imagination.

The focus in this chapter will be twofold. First, you will have the opportunity of choosing from twenty-three individual activities to develop your own creative spirit. A diversity of approaches is represented so you may choose those that keep you at your edge of growth. Although there are many more routes to enhance your imagination, these are a good start. Second, you will explore how other people may join with you to enrich your collective creative reasoning. Depending on your approach, others can be a drag on your own innovative efforts or they can be a spark, a catalyst.

To intensify your imagination, you have to broaden and

deepen yourself, as well as seek additional avenues of discovery. The ideas presented here augment suggestions sprinkled liberally throughout Chapters 3 and 6. To enhance and facilitate groups, you have to stimulate synergy or group genius and keep meetings on the creative cycle track. These suggestions build on and enlarge the chapters which describe the creative cycle, Chapters 4 through 9.

THE INDIVIDUAL'S GRAB BAG

Reach into this bag of activities to enhance your individual creative thinking. Which ones have you tried? Which ones make you excited or uncomfortable? Remember that a sense of discomfort, strangeness, or silliness, in addition to or in place of excitement, often accompanies new learning.

1. *Fuel your memory and experience bank.* Read widely. Penetrate beyond your own field. Delve into fiction and nonfiction, drama and comedy, biographies and exposés. Experience a diverse range of cinema and video productions. Glance through a variety of professional and nontechnical trade journals, noting the different issues and methodologies presented, even advertisements.

   An imaginative Nike shoe company senior designer, Tinker Hatfield, has sage advice for his colleagues. He suggests that, despite their greatly compressed free time, they expand their mental palette. Hatfield encourages them to go to the mountains, plays, shows, museums, and concerts. When the workers apply their creativity, they will then have a wider range of tools and experiences from which to draw ideas.[1]

2. *Associate characteristics, intuit solutions.* Choose a random word from the dictionary, an Edward de Bono technique. Or choose a neutral word, like tire, submarine sandwich, or rowboat. Freely associate attributes or characteristics of that word. Practice building solutions to a problem

based on those characteristics you just identified. Configure your responses in pictures, symbols, or words.

3. *Expand your consciousness.* Get in a soft focus state to search for ideas, clarify your thinking, and practice imaging solutions. Three interrelated tools will help you achieve a more open, imaginative level of consciousness. Conduct this soft focus exercise for 20 to 30 minutes one to two times a day, preferably twice. Start with relaxation, then begin to meditate and to create mental imagery.

- **Relax.** Sit easily in a comfortable arm chair with arms and legs uncrossed. With eyes closed, develop a relaxed state for 6 to 7 minutes using a method that works for you.

    One method of relaxing begins by paying attention to your breathing. Breathe in coolness. Breathe out frustration. Then tighten a muscle group for 5 to 7 seconds. Relax that body part until you feel no tension. Move to a new muscle group. Start with your dominant hand. Move to your forearm, upper arm, next hand, forearm, upper arm, top of head, face, neck, shoulder and chest, abdomen, pelvis, dominant thigh, calf, foot (stretch rather than tighten), next thigh, calf, foot.

    Since this will take more than 6 to 7 minutes, complete these body areas in groups, following a sequence over a period of days. Then tense muscle clusters (e.g., arms, head, upper body, lower body) when you have learned to relax completely and quickly.

- **Meditate.** Once relaxed, you can apply one of two meditation methods.

    One method is **focused.** *Explore specific problem or opportunity motifs and options using imagery and past experiences. Probe for new relationships and connections. When images appear spontaneously, you may hold and explore them. A second approach is* **undirected.**

> *In this method you clear your mind of all distractions. Be aware, nonjudgmentally, of sensations of your body and thoughts of your mind. Allow images to arise and disappear spontaneously without trying to control them. Receptivity is the key here.*

Capture ideas from the meditation period in your journal. As you write down your new thoughts, allow free associations to make further connections.

Visualize. You can expand your visualization abilities during meditation after the relaxation period by using focusing techniques. Do these regularly and make the images as vivid as possible.

> *In your mind's eye, create a chalkboard and, using chalk, draw each letter of the alphabet. Or picture a balloon floating in the air. Follow it wherever it goes. Make the balloon and its changing surroundings as brilliant and intense as possible, exploring both details and vistas.*

For further development of your image-making and meditative capacities, turn to Arnold Lazarus' book, *In the Mind's Eye: The Power of Imagery for Personal Enrichment* or Mike and Nancy Samuels' *Seeing with the Mind's Eye: The History, Techniques and Uses of Visualization*.

4. *Learn then practice aikido, an old martial art that can help build inner awareness.* Explore this body-mind process as a centering, broadening, and softening mechanism. Get an instructor more tuned to aikido's personal growth than its defensive aspects.
5. *Create creative space in your home, den, or office.* Hang pictures or dress shelves with artifacts that remind you of creativity. Strategically locate note pads to capture wisps of ideas as you work or play. Keep your journal with you so you reinforce your idea generation when you are on the move. Develop the habit of jotting down each brain-child.

One creative engineer keeps a stack of 5-inch-by-8-inch cards by his chair. He jots ideas, images, and solutions down on them, then squirrels them away in a logical place for that project. The cards allow him to conveniently deal with messages for several separate projects at the same time.[2]

6. *Develop a network of collaboration and support.* Seek out others who want a mutual exchange of ideas and perspectives. Include a mix of disciplines.
7. *Practice storming in your journal.* To flex your thinking, stagger the flow of ideas by generating first a common idea followed by a wild option. Then continue the common-wild sequence.
8. *Put the essence of what's said during a lecture or talk into images, pictures, or symbols, rather than taking notes with words.*
9. *Take fantasy trips.* Try this one that I came across in a stimulating workshop several years ago.

Close your eyes. Picture yourself taking a trip to see a wise old person about a problem or opportunity you have. You are in a mountain meadow. Before you take the trail into the woods, you pause for a moment to reflect about the problem or opportunity you face.

As you now head up the trail, you notice the trees, smells, wind, temperature, sounds, contours of the land. You continue on the trail for some time. Finally, deep in the woods, you arrive at a moss-covered cabin. Smoke is curling out of the chimney. A wise old person comes out of the cabin to welcome you, then takes you to the door and tells you that you'll find a solution to your problem inside the cabin.

You open the door and walk inside. Once in the cabin, you allow a solution to unfold around you. It may not make immediate sense, but you let it happen. Explore the solution image nonjudgmentally. You may decide to draw the image when you complete the journey. When ready, take the trip back to the meadow, slowly open your eyes, then sketch the solution image in your journal.

10. *Go to a workshop that will boost your applications of individual or group creative thinking.* Find a program that can enrich your emotional expressiveness, small group facilitation, or communication skills. Such workshops can help you listen more carefully to nuances of thinking, add emotional color that can spark a meeting, and become adroit at helping a group reach synergy. You may meet others who are also interested in creative thinking, possibly building friendships and new worlds to explore.

11. *Try your hand at artistic ventures that you might not ordinarily do.* Many inventive scientists and researchers have creative outlets away from work.

    Robert Root-Bernstein, a Los Angeles biochemist, studied many biographies of scientists. He concluded that most creative scientists have an artistic bent—painting, writing poetry or fiction, playing a musical instrument. Art and science, he says, are unified in the creative process.[3]

    Therefore, expand your modes of expression by keeping artistic channels open. Carve totems, play the piano, paint landscapes, work with wood, write novels, make pottery, custom design jewelry.

12. *Reach into the garbage can of ideas and breathe new life into them.*

    Deliberately take ideas that don't seem feasible or are too wild or seem to be off the mark. Discover the possibilities in these discards. Storm ideas to help them become winners.

13. *Find new uses for real castoffs.* Rummage through junk yards, thrift stores, or flea markets. See what you can do with real discards or hand-me-downs. Rube Goldberg could have a field day in such places, as can other scavengers, artists, and inventors. Let your imagination roam, turning a motor with a worn drive into a chisel sharpener, an old door into a hanging desk, odd gadgets into light fixtures.

14. *Change your self-talk.* You, like everyone else, talk to yourself much of the time. Usually that internal conversation goes on automatically. You have been your own companion since birth and have lots of opinions about what you do well and what you mess up.

No, talking to yourself doesn't mean you have "gone bananas" or are "way out in left field." Sometimes, though, your self-talk can defeat your innovative thinking. You may mumble to yourself that something is too difficult for you or too silly or requires too much skill or will be judged too harshly or is too wild or too "out of character." When you talk to yourself like that, you may be blocking opportunities to use your creative energy. First, recognize your inner negative chatter, then deliberately reframe those put-downs into positive statements.

Think of some of the things you say to yourself that might stop you from fully using your imagination. Jot down some of those self-defeating statements.

A powerful method to change inhibiting behavior is to act "as if." Use affirming statements "as if" you were already thinking that way, ones that will lead and release imaginative behavior. Tell yourself, for instance:

- *With each passing moment, I am feeling more alive, more imaginative.*
- *Surely there is an exciting solution for this problem just waiting to be uncovered. I will be the one to discover it or to pave the way for others to unearth it.*
- *I allow myself to become increasingly relaxed, flexible, and soft focused, exploring new ideas like an inquisitive child.*
- *I don't worry that the solution doesn't make sense. Its true essence will emerge.*
- *Like creative persons before me, I let my imagination play with ideas, withholding judgment until the appropriate time.*

- *I am not afraid to think boldly or to strive for ideas that no one else would conceive, for I can readily tame my proposals without defensiveness.*

15. *Join tiger or "skunk-works" project teams.* These are high energy groups that often require imaginative, interdisciplinary approaches to special product or organizational problems.
16. *Look at problematic situations from different angles.*

    Reverse an approach. State your targets as doing the opposite of what you want, then tease out the characteristics that cause you to produce those reversed conditions. You'll have a new view of what you want to avoid. Contort situations or proposals. Rotate, stretch, shrink, massacre, round, square, flatten, multiply, or spindle them.

17. *Ask others what they do to stimulate their creative thinking.* Especially ask people who may be of a different psychological type than you.
18. *Strive for serendipity.* Accidental discoveries have been very important to science. Chance discovery could be left to chance. Or you can actively pursue serendipity.

    In the mid-1980s, Patrick Hannan, Rustum Roy, and John Christman cochaired a serendipity symposium. They alerted scientists to the role of chance in discovery, described numerous examples they had collected, and suggested things the scientific community could do to buttress the function of serendipity.[4,5]

    Taking cues from that trio, here are several things you could do:

    - **Prime your mind.** Louis Pasteur, who accidentally discovered new uses for vaccination, said that "chance favors a prepared mind." Tackle your work with a keen sense of observation. One of my former colleagues turned many mishaps or problems with his

experiments into patents. He was alert to what went wrong and sought the whys.
- **Doggedly follow leads.** Remember the story of the British veterinarian in the nineteenth century who once told his students to throw away bacteria cultures that became contaminated with a fungus. Early in this century, Alexander Fleming became curious when the same thing happened. He kept after the anomaly, eventually producing penicillin.
- **Support serendipity.** Join with colleagues in musing over peculiarities in your data. When reporting results, document unexpected findings. Describe the way it really happened, rather than scrubbing the findings too carefully for a sterile reporting in the journals.

19. *Dig a tunnel.* Or find your own hideaway where you can release your imagination.

> Seymour Cray, designer of the ultra-fast Cray computers, has a refuge to go to when he gets stumped. He goes to his tunnel, an 8-foot-by-4-foot happening being carved out underground at his Wisconsin home. Cray laughingly says there are elves in the woods that come out when he's digging in the tunnel, and they solve the problems he's facing.[6]
>
> Orlando A. Battista, a highly inquisitive man with over 500 patents, respects the realm beyond consciousness. Prominently placed in his office is a reclining chair. When facing a difficult problem, he takes a nap, allowing subconscious processes to help him find solutions.

Develop your own ritual and place where you are free to explore.

20. *Computer your way to creativity.* More and more computer enthusiasts are finding ways to have their computers usher them into the realm of imagination. There are growing numbers of software programs that propose to provide the route to creativity, having you follow

ENHANCEMENTS: ENRICHING INDIVIDUALS AND GROUPS 231

thought processes that mimic the inventive brain at work.

For Walter Wilson, head of Merlin Technologies Inc., and his team, designing a new bobsled for the U.S. Olympic team took many creative leaps by using a new software package. The program, "The Idea Generator," by Experience in Software Inc., caused them to carefully determine what they wanted to create, then helped them storm numerous creative ideas.[7]

But it doesn't have to be an esoteric new program. The quick feedback response of my word processing and graphics applications has given me a much wider range of creative expression. Others find outlining, database, CAD/CAM, electronic mail, and graphics programs stimulating, triggering new ways of observing and creating. Also, scientists and engineers who have access to the imaging capabilities of superfast computers provide stunning new images and insights. Those individuals are paving the way for major leaps in many fields.

21. *Tease the brain.* There are many mind-challenging puzzles and exercises to increase mental flexibility, forming of new connections, visualization, and novel approaches. Here, for instance, are a few mind teasers:

- Name all the things that bug you.
- What if your nose were on top of your head?
- Unravel the meaning of each of the following four puzzles:

1

keep it

2

friends STANDING friends
     miss

3

V i o l et

4

Answers: 1. Keep it under your hat. 2. Misunderstanding between friends. 3. Shrinking violet. 4. Stars in your eyes.

- Arrange ten trees into five rows with four trees in each row.

If you enjoy these teasers, there are several books that should stimulate you. Eugene Raudsepp has crammed seventy-five mind-stretching games into each of two books, *Creative Growth Games* and *More Creative Growth Games*. Karl Albrecht has laced his *Brain Power: Learn to Improve Your Thinking Skills* with numerous mental teasers. Roger von Oech delivers enjoyable, perceptive tips for stirring your pot of imagination in *A Whack on the Side of the Head* and *A Kick in the Seat of the Pants*.

22. *Pick from a plethora of products and processes*. Gadgets, tools, and helps for those who desire to be creative continue to emerge. There are many routes to expanded mental awareness and tapping intuition. Individuals differ in their receptiveness and ability to use or benefit from the wide variety of helps, from the different media and modes. Some of these tools are still being tested. Here are a few of the latest:

- **Machines that bathe the brain in an electromagnetic field or direct electrical current into the brain.** One machine, for instance, Alpha-Stim, has been used by Stanford's Biofeedback and Stress Reduction Clinic founder, Arnold Brauer, to reduce stress and induce relaxation. Used in a double-blind experiment, Alpha-Stim, which sends electrostimulation to the brain through the ear lobes, reportedly enhanced learning on a computer game. Researchers haven't pinpointed the mechanisms for such an effect.
- **Isolation meditation.** Although floating on salt water is the old fashioned variety, a new isolation enclosure, Star Chamber, is a 3-square-foot enclosure with nine mirrors. The mirrors and special sound and lighting effects attempt to usher users into an eyes-and mind-opening meditative experience.

- **Soft focus sounds.** Many new cassette tapes use selected baroque music, synthesizers, and exotic musical instruments to create an open-minded, relaxed mood. By directing selected combinations of music and sounds to different ears and parts of the brain, some people seem to increase their intuitive reasoning. Used alone or in conjunction with music, the sounds—ocean waves, beeps and hums, multiple layers of voices, echoes—seem to break the normal mode of perceiving.[8,9]

23. *Feed your brain to get the most from your mind.* The food you eat will help dictate the amount and kind of mental energy you have available throughout the day. Does your mind turn to mush in midmorning? Does irritability hamper your ability to concentrate in the afternoon? Observe your eating habits carefully. Use strategies to let the food you eat set the mental stage for you to be more creative and your mind to be in peak condition when you want it to be.

    Judith Wurtman, a researcher at M.I.T., can help you pause before you stick food into your mouth. In her Managing Your Mind and Mood Through Food, *she dishes up sage strategies for using food to control the level of mental awareness. Since food triggers chemical reactions in the brain, you need to pick the right food to set off the chemicals that will give you the most payoff for the mental level and mood you want. Here are a few thoughts to digest:*

    - *Eat foods with protein to become or stay alert. Fish, skinless chicken, and shellfish, for example, are excellent sources for stimulating the chemical reaction that will get the mind ready for vigorous mental activity. To retain a mental edge, dive into the protein on your plate before you tackle the carbohydrates.*
    - *Munch on carbohydrates to get calm, focused, and relaxed. A piece of pie, crackers, potatoes, or a candy*

234 CREATIVE THINKING AND PROBLEM SOLVING

> *bar—sugars and starches—will reduce jangled nerves. That's right! Sugars such as a candy bar, according to research, calm not induce energy. Taken for breakfast or lunch, however, carbohydrates can cause a mental slump in midmorning and cause you to fall asleep during the boss's speech in the afternoon. When you are stressed out late in the afternoon, though, a carbohydrate snack can help you relax and refocus. Such a snack before bedtime will help you sleep.*
>
> - *Drink coffee, tea, or a soft drink in moderation early in the morning to get caffeine for additional mental stimulation. Avoid the morning coffee break, for there should be enough caffeine in the body to last until another gentle caffeine boost at noon. But don't take drinks with caffeine if that isn't your normal habit.*
> - *Avoid alcohol, high fat content, and high calories when you want to retain full mental capacities. Alcohol will sedate. Fat and calories will deprive your brain of blood because the blood will have to work overtime digesting food in the digestive tract.*[10]

This grab bag contains a variety of approaches to enhance your creativity that can be added to the many opportunities offered in earlier chapters. Try tailoring a few of the foregoing activities to your needs. Push yourself beyond your usual media or routines. Then cultivate and enjoy these challenges to growth. Now look at ways to enhance the use of others in your creative pursuits.

TAP GROUP GENIUS

Playwright and screenwriter Sampson Raphaelson was interviewed several years ago on Bill Moyer's television series on creativity. Raphaelson looked back on his Hollywood and Broadway success and mused about the past. He said he would sometimes stunt his creativity because he would get jealous of his style and not always share the creative act with others. His ego would get in the way. When he was able to join with others to create something, the creation was more than either one could have imagined.

Individuals joining together to produce more than they could have alone! That is what *group genius* is all about—*individuals huddling together for synergy*. Group genius is the *serendipity that comes when people thinking together find that 1 + 1 equals not just 2, but 30, chartreuse, or siss boom bah!*

Some scientists and engineers, especially those who have a tendency toward introversion, distrust team action. By being more comfortable with concepts and proposals that arise within themselves, they miss the creative power of group genius. Separately burrowing along their own paths, would-be inventors may not have enough pieces of the solution puzzle to create breakthroughs in today's complex technological climate. By linking thoughts with others, however, they can stimulate novel combinations, directions, and insights.

One aspiring inventor with over fifty patents bearing his name, Charles Eichelberger, works at Schenectady's General Electric Research and Development Center. Eichelberger gets inspiration from Edison's pioneering work that bagged the famous inventor over 1000 patents. But he also knows that many others contributed ideas and experimentation to making those patents find a commercial foothold.

To Eichelberger, the problems are becoming so complex that a team is often required to produce worthwhile ideas today. Ideas tossed around over a cup of coffee or during an informal drop-in office visit with colleagues from other disciplines provide clues to some technical dilemmas.

While trying to find a way to expand computer memory, for example, he picked up a critical piece of the puzzle while listening to colleagues talking about radar. It was the link he was needing and resulted in his designing a wristwatch-sized television camera that could be used for guiding robots. Others at the center went on to refine his new product idea.[11]

Eichelberger's group genius sprung from informal conversation. Other teaming efforts, ones that are structured to generate ideas, also produce helpful clues or solutions to baffling problems.

Such was the case for a group of geologists with a task of thoroughly mapping a difficult landscape, including rugged canyons. Making accurate observations on rolling terrain proved tedious but not undoable. However, when they came to a series of jagged, difficult-to-climb cliffs, they realized they needed a new approach.

They put their heads together in a meeting to storm possible ways of measuring the unusual cliff contours so they could build a consistent, accurate map. The brainchild that made them excited and worked very well was to hire a helicopter and a photographer.

They obtained control of the process in two ways. One was to form a team of helicopter pilot, photographer, and geologist to assure accurate coverage of the cliffs. A composite could then be carefully made from enlarged photographs. A second means of controlling the scale was to climb up some of the canyons to make measurements that could then be compared to the map. The result was a novel approach, accurate observations, and excitement for the geological team.

When to Call on Group Genius

You want lively, creative, focused get-togethers to do real work. People don't care to sit in meetings uninvolved or unappreciated. They are often eager to contribute when they know their ideas count. Pull people together under the following circumstances:

- **When an individual, group, or organization has a problem or opportunity and needs fresh or original ideas.** Some people who have attended my workshops have gone back to their companies and gathered colleagues together to storm ideas for technical, organizational, or people-oriented problems. They sometimes become the client needing consultation and ideas. Or they stimulate the identification of engineering or research areas that need creative attention, then jump in to use their new creative thinking techniques.
- **When ownership or support of the solution is needed by more than the person who needs to come up with the**

idea. Inventions often require time to be nurtured into existence, then grown and shaped into useful products. Many people may need to give their blessing before the invention hits the market or becomes part of a manufacturing process. Invite people to a group genius session when proposals are actually needed. You will gain not only good options but the impetus and support to launch ideas into adulthood.

- **When the urgency to produce the idea isn't immediate.** You may feel you are constantly under pressure to produce solutions and don't have time to call on others for ideas. Organizations often go through fire drills, causing jumps to be made through hoops at Olympic speed. Customers bark loudly and want immediate corrections when something appears to be wrong with a product or service. Breakdowns occur in processes and assembly lines. Much of the time, however, a team could be pulled together to storm ideas, even under time constraints, and produce wiser resolutions than the solutions many individuals make because they feel the pressure of time.

Group Enhancements: Enriching Group Genius Facilitation

Group genius can flourish and provide rich, innovative dividends to those who imbibe in team creative thinking. What's needed to move toward group genius? You promote group synergy by having the right mix of people, an effective process, a freeing-up environment, and skilled facilitation.

In Chapter 1, you looked at who to have on your team. A nourishing environment is discussed in Chapter 11, "Breakthrough Environments: Setting the Climate." Several key process helps have already been presented—the creative thinking process, many search strategies that can be applied in groups, and communication tips that attended each phase of the creative cycle.

In this part of the chapter, I will describe two ways to enrich your facilitation of group genius. First, learn and use the forces of group dynamics. Second, develop a practical meeting design in which to use the creative cycle.

Know Group Dynamics

To be equipped to handle team creative thinking, you must know what happens when people form groups. Teams have much collective strength. Because the interplay of individuals produces a unique combination of forces, however, you have to be keenly observant of potential snags to be overcome, as well as possibilities for group genius. Groups are dynamic because of the following.

MORE DATA ARE AVAILABLE. Most teams don't come close to mining the wealth of information in the group, from a wide mixture of ideas to insights into how to make something work. One of the problems with this information is that it is tucked into different recesses of each individual's mind.

Two psychologists, Joseph Luft and Harry Ingham, developed a time-honored concept for knowing where this information within individuals and groups is housed. In their model, information is grouped into four categories:

- *Category 1 is data out in the open, available to be used. It is information known by all members of the group—a written report, words spoken in a meeting, ideas written on newsprint.*
- *Category 2 consists of blind spots, or data unknown to us but known to others—such items as our bad breath, our parental gestures and tone, or appreciation that others have for work we have done.*
- *Category 3, our facade, contains that information we keep hidden from others—our simmering over a colleague's putdown, key data we hold back, thoughts that we fear will be seen as too silly or be judged harshly.*
- *Category 4 is the unknown, part of the arena of interaction between team members that is unknown to all parties. This information could be a recently developed formula or newly uncovered finding not yet known to team members. These data could also be unconscious images, events, feelings, or goals that require prompting to trigger associations that bring them to consciousness.*[12]

How does information get to the open arena from the blind spot, facade, or unknown? It takes keen listening and observational skills, trusting give-and-take, and willingness to self-disclose and provide feedback.

GROUPS CONTAIN A POWERFUL MIXTURE OF STYLES AND ROLES. Here's where psychological type theory sheds some light. As described in Chapter 2, you measure your type by taking the Myers-Briggs Type Indicator or The Keirsey Temperament Sorter, the latter being available in the book *Please Understand Me.* There are sixteen different styles or combinations of preferences.

For those uninterested or not knowledgeable in the thought patterns of each style, facing a range of people types in a creative problem solving session could be uncomfortable, even formidable. Discounting differing personality approaches can cause group genius to evaporate.

Roles people play can also be a complicating factor. There are the more formal roles—project leader, manager, sponsor, individual contributor, expert. Spontaneous or less formal roles also have to be handled—clown, idea shooter-downer, smoother, compromiser, cattle prod.

Some teams get so homogenized and cohesive that differences become blurred. That situation occurs when groupthink enters in, when conformity doesn't allow group genius to get off the ground. Good facilitation and interactional skills, however, will use the types and roles to foment creative, nondisruptive spice.

TASK AND MAINTENANCE ISSUES INTERACT. There is a tendency for researchers and other technical experts to short-circuit the free flow of ideas in a storming session. They get caught up in a discussion of technical merits of ideas. The attraction to mechanics and techniques is like a log caught at the edge of a whirlwind or a moth being drawn ever closer to a fire. Scientists and engineers often need to be eased or weaned away from such discussion to assure time for an ample flow of new thoughts.

A group may become entangled in the content or the task to be performed. That entrapment may blind them to group member needs or how effectively the group is conducting its business.

Balance maintaining group satisfaction with doing the right things to complete the desired task. Gain facilitation skills to help you over this hump. Invite impartial individuals to meetings to help provide objectivity.

Many factors call for us to be alert and facilitative, to foster group genius, not groupthink: (1) Levels of data. (2) Preferences for perceiving, judging, handling the world around us, focusing our energy. (3) Informal and formal roles. (4) Tension between completing tasks and maintaining group needs. It is the interplay between these and other forces that makes groups dynamic.

Follow a Good Meeting Format

Enhancement of a meeting or a retreat starts with designing a good format for the session. There are sound steps to follow to have a productive, imagination-inducing meeting. They include the creative cycle, but are presented here in a meeting context, considering the practical flow of events.

If I am to lead or facilitate a meeting or retreat, I always start with a session design that incorporates the original intentions and hoped-for outcomes. As a meeting progresses, I will make subtle or substantial changes in the design on the spot. I want to be ready to meet the evolving needs of the group, new implications about the size of the issues being considered, and additional factors that emerge. The design is a flexible structure to provide a focus for immediate steps, as well as end goals. Here are the steps of one useful design:

1. *Develop expectations in a notice to participants before the meeting.* Create the mental frame of mind you want them to carry into the session. Tell them to come ready to turn on their imagination. Inform them of desired outcomes. But let them know that, as they put their heads together, the team might actually arrive at a different end point. Help them know that their input could make a difference. Let them see the get-together is not a do-all, end-all meeting. Frame the session as an opportunity to focus attention on areas important to the team without interruptions

of jangling telephones and drop-ins. Say that follow-up will be essential. Along with confirming the meeting time and place, let them know that casual clothes are appropriate.
2. *Start the meeting with a brief kickoff from the leader.* Review the general desires for the session. Describe any special roles to be taken by the leader and, if there is to be one, the facilitator.
3. *List expectations of participants on newsprint.* Get any personal agendas and needs on the table. If this is to be a retreat for your work team, you could ask, "We are going to spend three days together. What are the special things you would like to have happen here or to take home with you?" This step allows participants to take some ownership in the outcomes.
4. *Ask, then list, "What do we have to do to pull off these expectations?"* If they are stuck about what you want, you might suggest, "I'd like us to put down a few of the guidelines we should follow to make this a highly productive meeting for all of us. For example, I believe honesty and openness should be one of the rules, so we can fully know the real situation and develop imaginative, useful solutions." Developing this list helps the group set the norms for how they will act together.
5. *Give a mini-lecture on wholemind and body capabilities.* Suggest the elements needed for creative thinking.
6. *Present a brief overview of the creative cycle.* Then dig in. Move through the cycle a step at a time. The main section of this book details that part of the meeting process. When you get to the search phase, use the creative thinking strategy format you determined before the session. Have several available. For some of the techniques, of course, you will need to have on hand such things as pictures, paint, or blank brainwriting forms. Start with the strategy that most fits the maturity and readiness level of the group.
7. *Be true to the process.* The four phases provide balance. A group is helped by a strategy format that drives them

toward an end point while keeping their minds loose and that tantalizes their imagination while satisfying the need for workable solutions. But sometimes in the intensity of interaction, the group can become stuck, off on a tangent, forgetting the intention of an activity, or sagging with "after a heavy meal" blahs.

When the group becomes stuck, engage the members. Discuss where the team is in the creative cycle. Suggest group members quickly review their progress and where they need to be in the time remaining. Solicit their ideas about how to keep the process moving. The amount of time needed to handle the issue may have been misjudged. You may have to regroup at another time or expand the time to complete the process.

8. *Nail down at least a few key actions.* Since problems or opportunities are often more complex than originally conceived, the target, search, and check phases of the process may take more time than was intended in the design. This may be all right with the team as the situation, targets, and proposals unfold. Getting a clear picture of what has to be done can provide both realism and energy. If there is a squeeze of time in completing the action phase, though, make sure you at least have actions to keep the process going. Determine to have a follow-up meeting or a subgroup to review proposals.

9. *Critique the meeting.* Ask what the group did that was effective and what was ineffective. Rate the meeting effectiveness on a scale of 1 to 7, from low to high. Seek suggestions for improving the next meeting.

Sound facilitation of a creative problem solving or opportunity-mining group starts with fundamentals—handling the group dynamics and following a solid process that can guide you through the session.

If you are to be the group facilitator, remember that you are a channel, a vehicle for the team doing its work.

Facilitate, don't grandstand. Model the imaginative spirit—blending the arational and rational, softening your focus, releasing your creative child. Know the impact of your thinking style on others. Take a workshop where you can try out small group skills and get expert feedback. Ask a group you have facilitated how you can improve the effectiveness of your meeting skills.

The more you increase your facilitation skills, the more expert you will be in using the creative cycle, as a group leader, facilitator, or member.

ENHANCEMENT EPILOGUE

Individual and group enhancements are waiting to happen. If there is a mountain to climb, it is scaling the heights of your imagination. If there is a guru to be found, the teacher might be your own internal wisdom. Or the teaching could be done by colleagues and mentors on your team—giving you feedback, mirroring your strengths and limitations, providing grist against which to learn and grow. Hopefully the activities and processes suggested in this chapter whetted your desire to take a journey into the land of individual and group genius enhancements.

Now we must turn to the final link in operating with creative power. We need to explore our environment and how to use it to cause imagination sparks to fly.

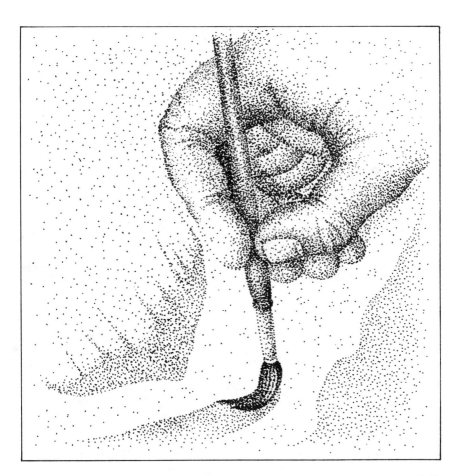

None of us lives in a lifeless, blank void. Forces around us can cause us to dull our thinking or to vibrate with creativity. We need not be bound in by tame surroundings. We can create our space. We can begin to paint our environmental canvas with colors and textures that stir and release the imaginative spirit.

11 Breakthrough Environments: Setting the Climate

> *We found, by trial and error, if everyone trusted each other, and you did not have a lot of control procedures, things worked much, much better.*
> —ROBERT SWIGGETT

> *...creativity may be kept alive in an atmosphere with minimal extrinsic constraint and maximal support of skill training, talent development, and intrinsic enjoyment of the work.*
> —TERESA AMABILE

Athletes, those hearty folks who tear around a field, whack or chase balls, and compete out-of-doors, often have an eye on the weather. How well they ply their sport may rest on how they respond to climatic conditions. Their muscles and skills become alert and responsive through hours of practice and conditioning. On the day of competition, the weather affects the way athletes play their game or run their race.

Perfect weather allows them to concentrate fully on performing to their utmost. They have sure footing, clear vision, and consistent conditions. Muscles and minds are responsive.

Poor weather may produce a whirlwind of reactions. Rain, cold, fog, wind, or snow can divert concentration, upset momentum, or disrupt the game plan. Finely tuned athletes can be turned into mock performers and the outcome can become a joke.

Your creative thinking efforts are similar to athletes on the day of competition. You also have an eye on the climatic conditions in which you try to exercise your imagination. You practice your discovery skills. You're ready to tackle a project that requires inventive thought. You want to hit upon breakthrough ideas, to run with an innovative proposal.

When the climate is perfect, you can focus your attention fully on producing a creative outcome. Your body, mind, and emotions purr, running relaxed and with a soft focus. If you are in a team, the group experiences synergy, producing exciting, challenging ideas. You trust the mental footing, appreciating everyone's contribution and knowing team members will back up the group-developed ideas.

An organizational environment conducive to creative thought will allow you to bounce ideas around with fervor and give your company imagination and potency. When the work climate is cloudy, stormy, or uncertain, your creative thinking can be disturbed. With an eye on the surrounding atmosphere, you may hesitate, not letting ideas run out of your mind with abandon. If you are worried about having your thoughts squelched, you'll be doing more protecting than generating. You won't be able to carry out your game plan with a sense of playfulness and risk taking. An enhanced organizational environment is essential if you want imagination and inventiveness to flourish.

ENVIRONMENTS BURSTING WITH CREATIVE ENERGY

You may have to take the lead in setting the right environment. If your feel lost, take heart. Help is on the way. Evidence continues to accumulate regarding principles, programs, and actions that build dynamic climates in which imagination is contagious. A string of researchers and observers of the innovation scene in America has recently produced exciting, sit-up-and-take-notice challenges to spark a creative fire in corporations. Their magnifying glasses have often focused on the corporate culture, the workplace climate that can be a creative firecracker or a dud.

Tom Peters and Bob Waterman struck a remarkably responsive chord around the globe, when their *In Search of Excellence* posed innovation as one of the marks of excellence. How to make environments in which excellence and imaginative thought are commonplace received much attention in Craig Hickman and Michael Silva's *Creating Excellence,* Tom Peters and Nancy Austin's *A Passion for Excellence,* and Tom Peters' *Thriving on Chaos.*

Rosabeth Moss Kanter's superb *The Change Masters: Innovation and Entrepreneurship in the American Corporation* describes organizations that cause ideas to wither and those companies that allow inventive thought to flourish. In *Innovation and Entrepreneurship: Practice and Principles,* Peter Drucker dips into his wealth of experience, study, and teaching. Once again he strikes a solid chord, showing how corporate planning and culture, along with management actions, can bear inventive fruit.

One of the books that gets to the heart of what turns creative thinking on or off is Teresa Amabile's *The Social Psychology of Creativity.* In it she presents her seminal research on the power of intrinsic motivation to facilitate creative thought and of extrinsic motivation to inhibit imagination.

People who find a task internally satisfying and challenging, she found, are generally able to be more creative than individuals who are trying to apply their imagination to meet an external reward or standard. Extrinsic motivation, such as feeling driven to meet a company award or a major milestone, can aid performance when the work is routine.

When real creative thinking is needed, intrinsic motivation is the key. Amabile's tests showed that extrinsic motivators could even erode the imagination of those who were known for their creative gifts.[1,2]

Such is the power of the work climate!

If we want to unleash that imaginative energy, we need to make our work environments a positive power for releasing innovative thinking. Research and empirical data, such as that described in the above books, provide clues. You may have good observations, too. I'll add mine.

Organizations, teams, and individuals can heighten the imaginative force of their work climate. Here are some suggestions to help make your environment crackle with creative energy.

Build Trust and Freedom

As Pete, manager of a life science unit, started the meeting, a few "butterflies" fluttered nervously in his stomach. Not only was he a new manager of a group of smart but sometimes caustic young scientists and technicians, but he was making his first attempt at facilitating a creative problem solving session. At first the group was suspicious of the process he was using, but they began to be encouraged by the techniques. They became energized as they analyzed the critical issues involved in trying to develop experimental techniques for a hot project of the laboratory director. Pete hoped they could develop imaginative new tests that could really get the project off the ground.

Just then Bill, Pete's director, burst into the room. In front of the team, Bill heatedly insisted that Pete remedy what he perceived to be a violation of one of the company procedures by Pete's group. Bill bolted out of the room, leaving Pete chagrined and his team bewildered. The team limped through the rest of the session, and came up with a decent but unoriginal experimental design.

This vignette is indicative of the type of disruptive climate in which individuals and teams sometimes try to be creative. Sound ideas may even result, but the imaginative flair is absent. The environment is too toxic.

At least two very important elements were torpedoed as a result of Bill's hit-and-run intrusion—trust and motivation. Trust was knocked off-kilter, causing suspicion and stifling the flow of ideas. Internal motivation—simply enjoying the challenge and synergism of the team freely coming up with ideas—grew, then took a nose dive.

An early organization development mentor of mine, William Owen, had a formula he liked to use in making teams and organizations more effective. Increased trust, he would say, leads to increased data flow, which in turn leads to increased enabling decisions. Improved trust, data flow, and enabling decisions equal improved productivity. Those same ingredients are crucial to allowing the imaginative spirit to blossom.

The key is trust, that critical attitude among people that climbs from skeptical to quizzical to confident to trusting. A restrictive environment causes us to easily slip back down toward distrust. Unfortunately, many organizations have a skeptical, quizzical atmosphere, snuffing out much creative thought before it can be kindled.

To affect the trust level, you have to be alert to what happens to you and others along the path from mistrust to full trust. Here are some descriptions and behaviors of the four trust steps, starting with the low end.

Skeptical

Individuals or groups are distrusted. Ideas they set forth are not felt to be worthwhile. Their proposals and suggestions are challenged. While the challenge could come as a hostile outburst, it is often couched in sophisticated put-downs. Mistakes are taboo. Humor is often a veiled attack, dripping in sarcasm. Good ideas are often kept to oneself, for fear of ridicule or having the flaws pointed out. Proposals are guarded and tame, fitting the receiver's bias rather than showing spontaneity and boldness.

People can be mistrustful and doubting unconsciously. Often the skeptical are so because of a stereotype or perceived slight or put-down by another. In many work cultures, individuals are expected to receive criticism without flinching, to blurt out their judgments and probe for flaws. In this faultfinding atmosphere, people may wear a veneer of acceptance and being a good sport, yet not be aware of the strength of their negative feelings and reactions. Even authentic, mature people can be hurt, fearful, and protective.

At this lower end of the trust continuum, freedom of the individual or group to think and act alone is curtailed. There is little elbowroom for taking chances. Controls control controls. Procedures define the range of thinking allowed. Motivation comes from outside the individual—from well-monitored reward systems to "crack the whip" enforcement behavior and policies.

Quizzical

Individuals and other work units are treated with a mixture of reactions. They and their ideas are treated cautiously—partly okay, suspiciously not okay. They may be seen as all right in a very narrow band, for example, if they stick entirely to their discipline and don't intrude into someone else's field.

The flow of ideas at this trust level is still distorted and incomplete, for evaluation is quick and judgmental. Questions abound—to test, probe, and provide some distance. Questioners seem to be saying, as a preface to their questions, "I want to know if your idea is safe, doesn't threaten the bounds of what I know to be true, and fits my knowledge, discipline, and experience. I want to make sure you know what you are talking about." Non-risky, more commonplace ideas may be acceptable.

Confident

People and their ideas are okay, even enjoyed. Communication and interaction begin to flow. Ideas are accepted as long as suggestions individuals make don't stray outside their area of expertise and experience. There is more freedom here to play with ideas. Risks of imagination may be taken by proposing options that are a bit strange or off-the-wall.

Trusting

People feel good about themselves and those they are around. Because individuals and groups feel honored and safe, any emotions, thoughts, ideas, attitudes, or behaviors are treated with respect. Crazy images, bizarre analogies, spontaneous humor, or wild combinations become natural responses of trusting parties engaged in creative discovery.

At this end of the continuum of trust, control is self-managed. This doesn't mean willy-nilly anarchy. Expectations are set and individuals and groups accept responsibility for high quality output in a spirit of dialogue and shared understanding. Then people are free to do their own work, free to be turned on by the challenge of the task, free to take risks and make decisions without heavy supervisory surveillance.

Companies that promote trust and discourage fear are able to release people, free them up to unleash more creative energy. One of the most important teachers and researchers of the role of trust in individuals and organizations is Jack Gibb. In his *Trust: A New View of Personal and Organizational Development*, he wrote:

> *"Trust is a releasing process.* It frees my creativity, allows me to focus my energy on creating and discovering rather than on defending. It releases my courage. It is my courage. It opens my processes, so that I can play, feel, enjoy, get angry, experience my pain, be who I am. The full life is spontaneous, unconstrained, flowing, trusting life.... When I get into my flow, all of my processes are heightened: my energy and breathing, creativity and imagery, awareness of sights and sounds and smells, courage. I become available to me and to others."[3]

Kollmorgen Incorporated discovered that a trusting attitude could help make it more successful. Chairman of the Board Robert Swiggett described the Kollmorgen story in Creativity Week IX, 1986, *a special report from the Center for Creative Leadership at Greensboro, North Carolina. In the 1970s, after acquiring a variety of small companies, each with a technological niche, Kollmorgen needed to find out what kind of a merged company it was and how to maintain an inventive edge.*

Swiggett commented, "When you get big you lose trust, you lose informality, you build up elaborate control systems. You build empires which have to be protected. You lose desire and vigor because the company gets kind of fat and successful. You lose the perceived necessity to bet the company on a new idea."[4] Building trust, therefore, was one of the key ingredients the company saw as essential for forging an imaginative spirit.

How do you build trust and freedom? Here are a few possibilities:

- At all levels of the organization promote managers who see people as trustworthy.

- Monitor where you, the company, or your team are on the trust hierarchy. Awareness is the forerunner of positive behavioral action. Get straight-shooting feedback regarding the accepting and rejecting behavior exhibited by you or others. The information will help pinpoint changes that need to be made.
- Be truly interested in the ideas others have. Show your acceptance of the creative contributions of others by inviting participation. Set up some type of quality of worklife or quality circle program. Empower others by pushing decisionmaking to the lowest levels possible.
- Paraphrase the proposals of others to seek an understanding of them. Honor the essence of the thought. Don't ask for ideas and then pounce on those that you don't like. Generate options to help fix problems inherent in suggestions.
- Use mistakes as learning tools.
- Reward those who have been trustworthy.
- Encourage people who have a creative bent. Accept zany brainstormed ideas. Laugh along with them as they try humorous attempts to look at something in a fresh way. Help them channel wild, boundary-pushing ideas into something that can be effective.

The dividends for high levels of trust are enormous.

Put Mouth and Actions Together

In preparation for the annual stockholders' meeting, the public relations staff huddled with the chief executive officer. They searched for the appropriate words to impress the public with the accomplishments and philosophy of the company, as well as convey the challenges that lay ahead.

As usual, the annual report would be a glossy centerpiece distributed to both stockholders and employees. Competition from other firms within and without the United States had caused their market share to shrink. Dear to the CEO were words such as "people are our most important asset" and "we have to continually strive for innovation in everything we do." He again wanted those thoughts woven into the report.

The ritual of the annual report is more than a public relations piece. It is a communication signal that can affect the culture positively or negatively. These often glossy reports can be omens about how close the company keeps its mouth attached to its actions. The values a company espouses in words may or may not be matched with deeds.

If innovation is to be a byword of your company, make sure you really want imaginative spirit and behavior. There are many driving forces in companies that may be at cross-purposes with an announcement that "creativity is honored here!" The forces include:

- Powerful people and organizational components
- Maintaining the edge of a superior product
- Keeping the company coffers full and stockholders happy
- Hefty regulations
- Chasing a "dream" that no longer fits nor has a champion or money to pursue it through lean and fat times
- Gyrating markets
- Internal as well as external competition

Examine the impact of these driving forces on developing and maintaining a culture in which the creative mind can exist through thick and thin. Test out the pluses and minuses, assets and liabilities that a bold statement about creative thinking might produce in your company.

In old-line companies, for instance, employees have a good memory for all the idea-squashing behavior in the past. If a philosophy statement about the importance of innovation receives chuckles or derision, there may already be a gap between what gets said and what actually happens.

If you investigate the climate, you might find anomalies to espoused company beliefs and behaviors, as I have many times within companies.

You might find that a basic, seemingly simple and straightforward technique such as brainstorming may be touted as a key to meeting effectiveness in a company. Yet the method can be bastardized and watered down because the ground rules for real storming go counter to the way individuals behave and are rewarded. Or the company says it wants your good ideas. But when some proposals are given, top managers laugh, ridicule, point out flaws, declare that the idea wouldn't fly here, or deprecate the idea's creator.

You can, however, make innovation be more than just a slogan. Breed it in as part of a living philosophy. Let resources, programs, and training be a natural outpouring of the philosophy. These natural products and efforts may include:

- Budgets for good ideas
- Seed money for the long haul of massaging an innovative thought into reality
- Tools like computers that can play "what if" games and develop scenarios
- Creative problem solving and opportunity development programs for individuals, groups, and organizations
- Training that prepares people to handle the risks and requirements of a fostering rather than festering organization.

The preparation of people to create an excellent, innovative climate is crucial. Hickman and Silva, in *Creating Excellence* describe the following six skills as necessary for this new age:

- **Creative insight:** Seeing to the heart of issues, asking questions perceptively, meditating to probe the essence of situations and possibilities.
- **Sensitivity:** Focusing on the real needs, feelings, and expectations of employees; sensing the uniqueness of each individual; allowing oneself to engage people at an intimate and trusting level; behaving in a way that exhibits true sensitivity.
- **Vision:** Creating the roadways into the future by ferreting through a plethora of data to arrive at a clear picture of

directions to take, finding the vision path through meditation and projecting scenarios of possibilities.
- **Versatility:** Broadening oneself into arenas that go beyond the immediate activities and goals, anticipating and acting upon the need to make changes.
- **Focus:** Directing attention on a few things rather than scattering efforts in multiple, resource and energy-sapping ways.
- **Patience:** Maintaining a perspective that can carry a person through the turmoil of the short-term, applying the skill of timing, developing staying power for following one's vision through the long haul to fruition.[5]

Without vision and well-thought-out words to express it, companies are hard pressed to back up innovation values. Actions reinforce the philosophy. Organizations, teams, and people, however, are open, dynamic, constantly changing systems. Therefore, the congruity between values and behaviors should be monitored constantly.

Model Creative Behavior

Barbara presented her value engineering staff with a knotty company problem that had plagued their chemical plant for years. She felt excited and challenged, describing to her team the need for innovative ideas despite the company doomsayers who predicted that nothing would come of their effort.

When Barbara asked for a storm of ideas and began to receive them, she quickly wrote them down on chart paper. When Ralph suggested a zany idea that regaled the team, she relished the craziness along with other group members and wrote the proposal on the flipchart, even though actually carrying it out might have banished her from management forever.

At the end of the 2-hour staff meeting, the team had chosen an approach, recognized the realities and possibilities of the challenge, and were charged up to find ways to solve the nagging problem. As the group tackled the issue, they met in subgroups—gathering data, holding sessions to storm subelements of the problem, enjoying the wild, yet solution-producing, energy of the team.

The moral of this story is simple but powerful. Staff members and teams take their cues from the leader. There is a conscious and unconscious eye trained on the manager, the facilitator, the one writing the paycheck and filling out the performance appraisal, the sponsor of the project, the big boss!

Leaders are often oblivious to how they model behavior and how followers mimic or key their actions to fit the modeled behavior. They are often not aware of how employees pick up habits, pet phrases and behaviors, casual thoughts. Followers fit the mold of the model.

When managers and other leaders become aware of idea-squashing behavior mixed in with their idea-generating behavior, they may plead, "Well, do what I say, not what I do." Awareness of the inconsistency is a good first start. The next step is to go for positive, proactive, creative behavior on a regular basis.

One of the most important, potent learning mechanisms is modeled behavior.

A friend of mine whom we will call Paul was head of a large public sector organization. He told me of a time during a meeting when one of the staff members insisted that someone had broken one of the organization's rules. When Paul asked where the rule had come from, he was dumbfounded to hear that he himself had set it. The staff member had observed Paul telling someone to change his behavior. The casual observation got translated throughout the organization as a rule that people were following.

A mimicking incident occurred at an IBM office, where chairs were colored gray and desks were made of steel. When a high-ranking manager brought in an orange ashtray, there were suddenly colorful objects cropping up on desks and in offices throughout that branch.[6]

Barbara modeled innovation-inducing behavior. If we were to dissect what she did, we would find many important ingredients that set an example for her staff. Her expectations for a creative outcome were translated into enthusiastic behaviors: a facial expression that mirrored excitement of the chase for novel ideas, words of encouragement, the fluid use of a creative thinking process in a staff meeting, a quick hand that recorded even off-the-wall thoughts without a flicker of judgment, and a readiness to keep pushing for innovative possibilities despite a few company catcalls.

Barbara legitimized the use of intuition, image making, approximation, and crazy ideas to stimulate thinking. She showed her support for such nonrational reasoning by asking questions sprinkled with creative prompts. "What does your gut say about what should be done?" "If your wildest fantasy about solving this problem were to come true, what would that be?" "We can always tame down an idea to make it fit the requirements later. What is something that possibly no one else would think of to solve this situation?" Her team picked up these imagination-spawning cues and began using these positive behaviors in subgroups and in their own thinking.

Modeling of behavior is one of the most natural, formidable forces shaping people's actions and attitudes. To encourage creative thought, therefore, model innovation stimulating behavior.

Build a Porous, Bendable Organization

Munching on popcorn and leaning back in a rustic chair, Zayre was ready for a couple of days of reflection. His first year as chief executive officer of a pharmaceutical company had been so hectic that he had had little time to do the broad thinking he felt was essential to the future of his company.

Now he was hidden away in a refreshing mountain retreat. Zayre, who had worked himself up through the company, knew he needed help to sort through his views of where the corporation was going and what would be needed to mold it further to bring out its full innovative luster. An organization development consultant was present to be a sounding board and to challenge the leader's thinking. Later Zayre would tap the thinking of his executive staff. Right now he wanted to get fully tuned in to his thoughts.

Zayre wants to have a vibrant, inventive organization and to provide the necessary leadership. Where can he get wisdom to guide his thinking? He, of course, may tap into his own wellspring of insight and imagination. In the hustle of work, that fund of vision and creative thought can become hazy or neglected.

He's chosen an excellent starting approach for probing his own thoughts. The retreat setting and third party assistance will help him gain perspective on his company and the role he plays in it.

As Zayre explores the creative environment of his organization, there is an additional fund of wisdom that can help him analyze and set targets. The books I mentioned earlier in this chapter provide a treasure chest of research, experience, examples, and principles.

Examine my paraphrase of a theme that is woven through a number of those books that could act as a measuring stick or a divining rod to Zayre, to you, or to anyone truly interested in fostering an imaginative organization: *Build a porous, bendable organization.* "Porous" doesn't mean having a lax security system or being riddled with loopholes where unproductive employees can exist. "Bendable" shouldn't connote wishy-washy or chameleon-like.

Dynamic, innovative, entrepreneurial companies allow themselves to be open systems, where ideas and information flow across organizational boundaries. These companies flex their structure and systems to meet emerging opportunities or issues. Change is sanctioned and welcomed. Temporary forms and processes handle the ebb and flow of challenges. Reward and benefit systems provide room for individual choice. Policies and procedures are often lean and can be altered to meet changing needs.

To erect an organization that is porous and bendable, there are a number of action stances that could help:

1. *Clean out the pores, the channels of information and interaction that allow organizations to breathe with nonstatic, multiple-way communication.* Pores can be structural, cultural, interpersonal, individual, directional, and procedural.

    Obstructed organizational pores are what Rosabeth Moss Kanter, author of *Change Masters*, terms *segmentalism*. This is the pervasive setting up of organizational compartments. Problems and ideas get stuck in cubbyholes of hierarchy, specialists, departments, procedures, and levels. Because of these segments, organizations play it safe, dwell on the past and conformity, restrict the flow of information, protect themselves from change, keep problems localized.

Kanter sees innovative organizations, on the other hand, as *integrative*. These organizations vigorously set out to eliminate roadblocks, clogged pores. To bring about this integration, information and interaction are allowed to flow across boundaries. Disciplines and viewpoints intermingle. A common vision is pursued. The future is trusted, a target for investments. There is healthy teamwork and respect for individual competencies. Issues and projects are tackled with common effort and vigor.[7]

To manage this integration, seen as chaos by some people, there must be the recognition that the energy spawned by this type of organization can bring creative conflict. Skills must be mastered for handling change and diversity of thought.

2. *Enhance flexibility and synergy by creating teams.* There is increasing recognition that numerous important projects require the dedication and synergy of teams. The complexity of problems and diversity of issues often demand the use of teamwork for effective solutions. Teams can take many forms. Here are a few exciting formats: "tiger" or "skunk works" teams; the Kollmorgen small team concept; "innovation" offices or teams; and organization development teams, such as quality circles.

Tiger or Skunk Works Teams

"Tiger" or "skunk works" teams, the latter term made popular by Tom Peters and Robert Waterman, are high energy, sharp-focus project groups. They are often pulled together for such things as developing a new product or process, unscrambling a difficult organizational problem, or producing a major proposal. In effective, innovative organizations, teams are alive and action-oriented, not stuffy, monotonous committee drones. These teams are at ease talking and interacting across disciplines.

One such group was set up at Data General's Westborough complex. Under the driving force of Tom West, a team of mostly young and talented computer engineers was pulled together to develop a new computer. Bright computer graduates were recruited. They were tantalized by the opportunity to leap into a challenging aspect of computer development, rather than to be relegated to a beginner's support role in some other company.

The work was intense. Time melted away, as tasks required much overtime. Obtaining resources and support from other quarters in the company sometimes required bargaining. Brilliant flashes were required of individuals. Coordinated teamwork was also essential. Tracy Kidder traced the team's breakthrough success in his Soul of a New Machine.[8]

Kollmorgen Small Team Concept

Kollmorgen used the small team concept another way.

The company discovered that innovation required a small team atmosphere, where people could be known personally by the head of a division and feel a part of the whole operation. Therefore, they kept cost centers to no more than 500 employees.

Innovation Offices or Teams

One of the creative forms of teamwork can be seen in the growing number of innovation offices or teams. These groups are often ad hoc task forces or standing committees. Their scope may include spotting such targets as uncertainties in markets, processes, and products; solving problems and generating ideas; evaluating new technologies; implementing proposals; forecasting; and selecting product suppliers that maintain an inventive edge.

Such diverse companies as Ford Motor Company, Battelle Memorial Institute, and Eastman Kodak have set up such teams or offices. Eastman Kodak, for instance, set up a system that calls for peer evaluation of an idea before management takes action.

Organization Development Teams

Another highly potent group format uses teams to increase productivity, raise output quality, renew the organization, improve life in the workplace, induce creative thinking. Branded by such names as product improvement teams, employee involvement, quality of worklife, or quality circles, these are organization development interventions. Their purpose is to bring about change within the whole organization or within specific targets of opportunity.

Such dissimilar companies as RCA, Honeywell, General Motors, Hughes Aircraft, Rockwell, and Westinghouse have found positive results with similar team processes.

Some managers naively try to ride this wave. They get swamped, however, because they haven't developed the program, support, skill, and resources to maintain these processes over the long haul. It would be like surfing for the first time with one hand tied behind your back and a blindfold over your eyes. Typically, these team efforts require:

- Commitment by managers at all levels in the organization, especially rock-solid support at the top
- A programmatic approach that can take a big picture view—assuring resources, training, monitoring of the process, and consultation by organizational development specialists knowledgeable in change processes
- Managers and employees willing to explore the hindrances to effective work and the impacts of their behavior on each other and the company culture
- Special techniques for identification and solving of problems
- Training in creative problem solving skills
- Intact work units or voluntary groups becoming teams to affect problems faced in work activities
- Skilled facilitators to coach teams and managers and to lead groups through the meeting process when that's appropriate
- Persistence—treating failures, resistance, and roadblocks as chances to learn and to refine the process over time
- Making these processes useful management tools and a way of life

These team efforts have great potential for making organizations exciting, imagination-inducing workplaces, as well as more effective and productive.

To get these results, have a zest for seeing other people become involved and challenged. Be committed to hearing the flaws as well as the positives. Be willing to model the behavior required for an open, flexible organization.

Zayre, the new chief executive officer in the vignette at the beginning of this section, can ponder the future directions of his company in his retreat hideaway with firmer knowledge. He now has spread in front of him many possibilities for reviewing and renewing his organization. They are the same means of creating porous, bendable organizations available to you.

Stimulate Mind and Senses

Winter's early morning darkness and frigid air provided an eerie blanket for a bedroom community 12 miles from the big city. Thousands of people were in the final delights of slumber. Then alarms began to ring and radios and coffee makers automatically came alive at prearranged times. As people stumbled out of bed, they varied considerably in their eagerness for rushing off and plunging to work in offices, labs, factories, or fields.

The moods of people going through the getting-up ritual could be symbolic of the type of stimulation that awaits them at their work location. Granted, there are factors other than the work environment that color an individual's morning mood—day-night rhythms, getting hooked on a late-night movie, a quarrel with one's spouse, late partying, or a host of other events or situations.

But we are drawn toward environments that have the right mixture of stimulation and avoid those that bore or jar us. Getting up in the morning to go to a job that is causing our minds and bodies to rust out or to burn up is not motivating. We can hardly turn off an alarm, brush our teeth, or get dressed, let alone tackle work with eagerness and imagination.

Albert Mehrabian provides illuminating concepts about our environments in his *Public Places and Private Spaces: The Psychology of Work, Play, and Living Environments*. Each environmental setting gives us a measurable amount of novelty and complexity. A drab, plain, uncluttered office, for instance, probably has little that is very new or diverse. A company picnic teeming with families and numerous booths and games, and with top managers acting unusually zany, poses considerably more information to process.

How we behave in our environment, says Mehrabian, is a result of what's happening to us on three emotional dimensions—arousal, pleasure, and dominance. Based on how we are feeling in these areas, we may want to move toward a particular environment or we may want to pull away or avoid it.

- Arousal, our level of excitement and alertness, can be observed in breathing and blood pressure rates, muscle tone, and readiness to cope with events. Response to this dimension is often the most potent of the three.
- Pleasure, how satisfied or happy we feel, can be noted in our smiles or frowns, tone of voice, and words we express.
- Dominance is the degree to which we feel in control or influential, guided or cowed. It is seen in our acting freely or in curtailing behavior to fit the "etiquette" of the setting.[9]

Although each of these emotional dimensions operates independently, they can have a combined effect.

How might a technical expert react to a situation in which she is being reprimanded by her manager for bootlegging material and taking company time to develop an idea the expert thinks is promising? High arousal might cause the jitters. Pleasure is certainly low, as well as the feeling that she can influence the situation. The combined effect would likely be a desire to avoid her manager or not take creative risks again.

To promote creative thinking, therefore, you want to develop an environment in which people are aroused, find pleasure, and have a sense of being in charge of their work and its outcome.

You especially need to assure that the mind and senses receive an appropriate level of stimulation.

Here are some things you can do to inject positive stimulation into the work environment:

1. *Make work challenging.* People are turned on by jobs that are meaningful, that make a contribution, that seem to make a difference. Employees want to stretch, to expand skills and knowledge.

   > *The heroes of* Soul of a New Machine, *whether fresh from school or weathered by experience, drove themselves and logged many extra hours because they were caught up in the excitement of creating a new computer. They felt their creation could make a difference.*

   People like to know they are growing and advancing, not being stagnant or stuck off to the side. But motivations differ. For one person the stretch could come from leading a project. Another would like to unravel a technical problem that has perplexed colleagues. Still another is challenged, even though a little scared, by developing a strategy and presentation for selling an idea to upper management.

   People need to know how they are contributing to the goals of the company. Whenever possible, employees want to participate in setting the aims and objectives. In rapidly changing work environments, it's extremely important to communicate shifts of direction and emphasis, allowing people to make adjustments in their work to fit altered priorities. On the other hand, employees don't like to be jerked from one project to another without ever feeling they can finish a product or project. It's hard to enthusiastically roll out of bed in the morning to tackle a third-rate project that may be terminated tomorrow!

2. *Assure the resources to do the job.* Give employees the equipment, technical support, material, time, and information to do their work. The remarkable story of the development of Post-it Notes has highlighted the fertile environment within 3M for producing new product ideas.

> Employees at 3M can take 15 percent of their work week to conjure up and chase ideas. Despite the vast numbers of people working at their sprawling St. Paul campus, individuals are often able to link up with others to exchange ideas.
>
> Art Fry, the individual who conceived the note pad concept and carried it through to its highly successful marketing conclusion, had a number of avenues open to him for gathering information. He had access to the weak adhesive developed by Spencer Silver, a 3M colleague. At his computer, Fry could access the bank of technical papers written by other employees. Or he could go to one of eleven libraries. Coffee breaks, the bane of efficiency experts, provided him with invaluable company and technical information. When the project was small and underfunded, Fry found materials he needed arriving from a variety of unofficial sources.[10]

Like any large, open system, 3M continually needs to tend to its culture to keep it vibrant, thirsty for new ideas, and responsive. The company has scores of new proposals in the works at any one time. Thus, individuals who want special attention given to their projects have to know the system and to find ways to make it work. Resources have to be available so employees are stimulated, not stymied. And managers need to encourage employees to use the resources.

3. *Grow entrepreneurs and champions.* Nourish those individuals who are willing to champion causes, to conceive solutions and be the vanguard for their inventions. Champions don't have to start with a self-imagined proposal, but just be very supportive of a proposal and add their leadership to turn the idea into reality.

> Entrepreneur Art Fry had a champion in Geoff Nicholson, his lab director. Nicholson took the note pads to 3M's executive offices and distributed them to the secretaries, allowing them to get addicted to the trial product.

> Soon the vice-presidents were caught up in the yellow-stickee craze and, thus, more amenable to sales pitches.

Homegrown entrepreneurs and champions, sometimes the same person, are like blood transfusions. These special people pump a fresh current of ideas through the company to vitalize both employees and the enterprise.

> Allow entrepreneurs to blossom. Relish the champion. Tolerate the maverick spirit. Reduce organizational barriers that prevent these human catalysts from taking fruitful risks.

4. Find vehicles to tantalize, encourage, and arouse the whole person.

> Neuroanatomist Marian Diamond from the University of California, Berkeley, has stated, "The environment can change the brain, and it can do so at any age." Diamond makes this bold statement as a result of studies she and others have done. The researcher and her colleagues have shown that rats placed in an enriched environment produce offspring that have thickened cortices. Kept in an enhanced climate, generations of rats produce ever-thickening cortices.
>
> Researchers in the United States, Japan, and Sweden are trying different combinations of physical environments to entertain the rats. Cortical growth doesn't come from using just one toy, but from rats playing in their cages with a variety of forms and levels.[11]

Much further research will have to be done to show how enriched environments can expand human cortices. Yet there are other exciting things happening at a variety of places around the world that show the potential for developing the mind. Improvements are being made in remembering things, athletic performance, and learning.

Combinations of baroque music, relaxation exercises, visualization techniques, and rhythmic intonations have been used to create an atmosphere to stimulate improved mental performance. Sheila Ostrander, Lynn Schroeder, and Nancy Ostrander describe such research, techniques, and results in their *Superlearning*.

Look for combinations of elements to provide a growth- and imagination-enhancing environment. Arouse and bring pleasure to the whole person with stimulating images, activities, textures, shapes, surroundings, appointments, interaction, equipment, and colors.

Herman Miller Research Corporation produced a book full of many suggestions for transforming the workplace into an arena that compels rather than repels. In their book, *Everybody's Business: A Fund of Retrievable Ideas for Humanizing Life in the Office,* the Herman Miller staff highlighted several companies they researched who seem to have created a more enriched environment.

One of the firms they spotlighted was Odetics, Inc., of Anaheim, California, a firm specializing in such esoteric high technology as robotics and space shuttle tape recording systems. Although much hard work and care goes into their products, the company tries to energize their environment.

Odetics is housed in esthetically pleasing facilities, with a well-equipped and heavily-used pool and fitness center. They even have a fun committee to cook up such ideas as lip sync contests and whale-watching parties. The Odetic Repertory Theater provides an outlet for the thespians in the company. The firm compares itself to a family, listening to both work and personal concerns.[12]

Add salt and pepper to your organization. Try a series of lunchtime talks and discussions. I was invited to present some creative thinking ideas at Crown Zellerbach's corporate office in San Francisco. They had a series of presentations dealing with fitness of body and mind.

Or have hobby and special interest clubs where employees can mingle, exchange ideas, and express their creative flair away from the job. Place art work, paintings and sculptures, around the workplace. Use many natural ingredients—plants, pools with fish, waterfalls, garden walkways, aviaries. Create incubation rooms with quiet areas or places with music, paint and crayons, paper, humor, and a chance to think boldly and visually.

There are countless possibilities for creating an enriched environment. You, and other kindred spirits, just need to use your imagination and, in some company cultures, your marketing skills. When you provide the right combination of stimulating elements, productivity and innovative thinking will take a leap forward. And reaching from under the covers to turn off the alarm and start a new workday will be considerably easier.

ENVIRONMENTAL EPILOGUE

Blockbuster, breakthrough environments don't just happen. They require vision and dedication to foster imagination.

You need to trade zipper-tight controls for trust and self-managed freedom. Make sure the company rhetoric has deeds to match. Dump the slogan, "Oh, well, at least do what I say, not what I do." Make a conscious effort to be an effective model of creative thought and action. Seek a porous, bendable organization, one that allows communication to flow freely and flexibly in structural joints. Keep the mind and senses alert, satisfied, and in control with a host of job-centered, environmental, and cross-fertilizing stimuli.

With a properly aroused organization, the creative thinking processes proposed earlier in the book will have a natural opportunity to produce breakthrough possibilities. From such an environment, you may well find the future imaginative leaders of your company emerging.

Notes

Chapter 1 The Quest: Pursuing Creativity and Innovation

1. Koestler, A. *The Act of Creation* (New York: The Macmillan Company, 1964).
2. Zukav, G. *The Dancing Wu Li Masters: An Overview of the New Physics* (New York: Morrow, 1979).
3. Furnas, C. C., J. McCarthy, and Editors of Time-Life Books. *The Engineer* (New York: Time-Life Books, 1972), pp. 40–53.
4. "Breakthrough Ideas: How to Be an Entrepreneurial Einstein," *Success* (October 1987), pp. 46–53.
5. Furnas, C. C., J. McCarthy, and Editors of Time-Life Books. *The Engineer* (New York: Time-Life Books, 1972), pp. 10–12.
6. Hampden-Turner, C. *Maps of the Mind* (New York: The Macmillan Company, 1982), p. 100.
7. "Frog's Gift to Man," *U.S. News & World Report* (August 10, 1987).
8. Kidder, T. *The Soul of a New Machine* (New York: Avon Books, 1981).
9. Imai, M. *KAIZEN: The Key to Japan's Competitive Success* (New York: Random House, 1986).
10. Ouchi, W. *Theory Z: How American Business Can Meet the Japanese Challenge* (Reading, MA: Addison-Wesley Publishing Company, 1981).
11. "Breeding New Ideas," *Newsweek* (August 8, 1988).
12. Gardner, J. W. *Self-Renewal: The Individual and the Innovative Society* (New York: W.W. Norton & Company, 1981), p. 31.
13. "Innovators: 100 Who've Made a Difference," *The Evergreen State Magazine* (Washington) 5(3):113–121 (November 1988).
14. Arieti, S. *Creativity: The Magic Synthesis* (New York: Basic Books, 1976).
15. "The Ups and Downs of Creativity," *Time* (October 8, 1984).
16. "Mood Disorders Affect 80% of Well-Known Writers," *Brain/Mind Bulletin* 13(3):1 (December 1987).
17. From Bill Moyers' 1982 television creativity series.
18. Maslow, A. H. *Creativity in Self-Actualizing People.*
19. From a 1981 Whole Brain Conference, Key West, Florida.
20. Khatena, J., and E. P. Torrance. *Thinking Creatively with Sounds and Words: Norms-Technical Manual*, research ed. (Lexington, MA: Personnel Press, 1973).

21. Samples, B. *The Metaphoric Mind: A Celebration of Creative Consciousness* (Reading, MA: Addison-Wesley, 1976), p. 100. (Used with permission.)
22. Bruner, J. S. *On Knowing: Essays for the Lefthand* (Cambridge, MA: Harvard University Press, 1963), p. 18.
23. Fox. *A Critique on Creativity in Science.*
24. Bailey, R. L. *Disciplined Creativity for Engineers* (Ann Arbor: Ann Arbor Science Publishers, Inc., 1978), p. 39.
25. Personal communication, 1989.
26. "Special Issue: Prigogine's Science of Becoming," *Brain/Mind Bulletin* 4(13):1–4 (May 21, 1979).

Chapter 2   The Human Instrument: Knowing Your Mental Equipment

1. Koestler, A. *The Act of Creation* (New York: The Macmillan Company, 1964), p. 217.
2. "Special Issue: Prigogine's Science of Becoming," *Brain/Mind Bulletin* 4(13):1–4 (May 21, 1979).
3. Jung, C. G. *Man and His Symbols* (New York: Doubleday, 1964).
4. "Special Issue: 'A New Science of Life,'" *Brain/Mind Bulletin* 6(13):1–4 (August 3, 1981).
5. Gleick, J. *Chaos: Making a New Science* (New York: Viking Penguin, Inc., 1987).
6. Keirsey, D., and M. Bates. *Please Understand Me: Character & Temperament Types* (Del Mar, CA: Prometheus Nemesis Book Company, 1984).
7. Fabian, J. P. "Creative Imagination Stimulation and Cognitive Style," Doctoral Dissertation, Washington State University, 1980.
8. "New Theory: Feelings Code, Organize Thinking," *Brain/Mind Bulletin* 7(6):1–2 (March 8, 1982).
9. "Creative Thought Emerges from Emotional Theme," *Brain/Mind Bulletin* 7(7):2 (March 29, 1982).
10. "'Confluent' Teaching Doubles Language Test Scores," *Brain/Mind Bulletin* 5(10):1,3 (April 7, 1980).
11. "'Optimal Performance' Similar in Athletics, Recovery," *Brain/Mind Bulletin* 5(8):1–2 (March 3, 1980).
12. Goldberg, P. *The Intuitive Edge: Understanding and Developing Intuition* (Los Angeles: J. P. Tarcher, 1983).
13. Koestler, A. *The Act of Creation* (New York: The Macmillan Company, 1964), pp. 141–144 and 170.
14. Garfield, P. *Creative Dreaming* (New York: Ballantine Books, 1974), p. 44. (Used with permission.)

15. "Dreams Led Barr to Pursue Melanin," *Brain/Mind Bulletin* 8(12/13):7 (July 11 and August 1, 1983).
16. Garfield, P. *Creative Dreaming* (New York: Ballantine Books, 1974), pp. 42–44.
17. Grinder, J., and R. Bandler. *The Structure of Magic II* (Palo Alto, CA: Science and Behavior Books, Inc., 1976), pp. 3–26.
18. Koestler, A. *The Act of Creation* (New York: The Macmillan Company, 1964), p. 170.
19. de Bono, E. *Lateral Thinking: Creativity Step by Step* (New York: Harper & Row, 1970).

Chapter 3   The Human Instrument: Expressing Breakthrough Qualities

1. From *The Dancing Wu Li Masters* by Gary Zukav. Copyright©1979 by Gary Zukav. Reprinted by permission of William Morrow & Co.
2. Fehmi, L., and G. Fritz. "Open Focus: The Attentional Foundation of Health and Well-Being," *Somatics* (Spring 1980): 24–30.
3. Furnas, C. C., J. McCarthy, and Editors of Time-Life Books. *The Engineer* (New York: Time-Life Books, 1972), p. 48.
4. *The Illustrated Science and Invention Encyclopedia*, international ed. (Westport, CT: H. S. Stuttman Inc., 1983), Vol. 4, p. 492.
5. Furnas, C. C., J. McCarthy, and Editors of Time-Life Books. *The Engineer* (New York: Time-Life Books, 1972), pp. 10–11.
6. *Open Focus: Cassette Training Series (Basic 1000)* (Princeton, NJ: Biofeedback Computers, 1977).
7. Zukav, G. *The Dancing Wu Li Masters: An Overview of the New Physics* (New York: Morrow, 1979), p. 117.
8. Brown, K. A. *Inventors at Work* (Redmond, WA: Tempus Books, 1988), pp. 1–18.
9. O'Neill, J. J. *Prodigal Genius: The Life of Nikola Tesla* (New York: Ives Washburn, Inc., 1945).

Chapter 5   The Target: Taking Aim

1. Poe, E. A. "Murders in the Rue Morgue," *Tales of Mystery and Imagination* (New York: The Mysterious Press, 1987), pp. 149–180.
2. Rossman, J. *Industrial Creativity: The Psychology of the Inventor* (New Hyde Park, NY: University Books, 1964).
3. Presidential Commission on the Space Shuttle *Challenger* Accident, W. P. Rogers, Chairman. *Report of the Presidential Commission on the*

Space Shuttle Challenger *Accident* (Washington, DC: U.S. Government Printing Office, 1986).

Chapter 6 Search Keys: Igniting the Imagination

1. Goodman, J., Ed. "The Fun Minute Manager," *Laughing Matters* 4(4):127–137.
2. Bailey, R. L. Personal communication, 1989.
3. Edwards, B. *Drawing on the Right Side of the Brain* (Los Angeles: J. P. Tarcher, 1979).
4. Koestler, A. *The Act of Creation* (New York: The Macmillan Company, 1964), p. 182.
5. Ray, M., and R. Myers. *Creativity in Business* (Garden City, NY: Doubleday, 1986), p. 6.
6. Brown, K. A. *Inventors at Work* (Redmond, WA: Tempus Books, 1988), pp. 45–67.
7. Brown, K. A. *Inventors at Work* (Redmond, WA: Tempus Books, 1988), pp. 19–44.
8. Zwicky, F. *Discovery, Invention, Research: Through the Morphological Approach* (Toronto: The Macmillan Company, 1969).

Chapter 7 Search: Following Steps and Strategy Formats

1. Presidential Commission on the Space Shuttle *Challenger* Accident, W. P. Rogers, Chairman. *Report of the Presidential Commission on the Space Shuttle* Challenger *Accident* (Washington, DC: U.S. Government Printing Office, 1986).
2. Buzan, T. *Use Both Sides of Your Brain*, rev. ed. (New York: E. P. Dutton, 1983).
3. Goldberg, P. *The Intuitive Edge: Understanding and Developing Intuition* (Los Angeles: J. P. Tarcher, 1983).
4. Gordon, W. J. J. *Synectics: The Development of Creative Capacity* (New York: Harper & Row, 1961).
5. Prince, G. M. *The Practice of Creativity: A Manual for Dynamic Group Problem-Solving* (New York: Collier Books, 1970).
6. Geschka, H., U. Reibnitz, and K. Storik. *Idea Generation Methods: Creative Solutions to Business and Technical Problems* (Columbus, OH: Battelle Memorial Institute, 1981).
7. Fox, R., R. Lippitt, and E. Schindler-Rainman. *Towards a Humane Society: Images of Potentiality* (Fairfax, VA: NTL Learning Resources Corporation, 1973).

Chapter 8 Check: Assessing Options

1. Presidential Commission on the Space Shuttle *Challenger* Accident, W. P. Rogers, Chairman. *Report of the Presidential Commission on the Space Shuttle* Challenger *Accident* (Washington, DC: U.S. Government Printing Office, 1986).

Chapter 9 Action: Launching Ideas

1. "Not Doing It the Company Way," *Success* (January/February 1988).
2. "Feats of Inspiration and Originality" *Time* (October 26, 1987).
3. "The Oops Factor," *Omni* 10(5):31 (February 1988).
4. Furnas, C. C., J. McCarthy, and Editors of Time-Life Books. *The Engineer* (New York: Time-Life Books, 1972).
5. Brown, K. A. *Inventors at Work* (Redmond, WA: Tempus Books, 1988), pp. 147–165.
6. Bailey, R. Personal communication, 1989.
7. Bliss, M. *The Discovery of Insulin* (Chicago: University of Chicago Press, 1982).
8. Brown, K. A. *Inventors at Work* (Redmond, WA: Tempus Books, 1988), pp. 335–351.

Chapter 10 Enhancements: Enriching Individuals and Groups

1. "Nike: Designers, Researchers Encouraged to Pursue Their Ideas," *Oregonian* (March 12, 1989), p. L9.
2. Bailey, R. Personal communication, 1989.
3. "Arts Enhance Scientific Intelligence, Study Says," *Brain/Mind Bulletin* 10(12):1–2 (July 8, 1985).
4. "The Oops Factor," *Omni* 10(5):31 (February 1988).
5. "Serendipity in Science: Survey Gathers Evidence," *Brain/Mind Bulletin* 11(17):1–3 (October 20, 1986).
6. "Fast and Smart," *Time* (March 28, 1988).
7. "The Idea Generator," *Success* (September 1988).
8. "A New Look at Brain Machines: What the Experts Say," *The Omni Wholemind Newsletter* 1(6):1 (May 1988).
9. Colligan, D. "Mindware Tools for Awareness," *The Omni Wholemind Newsletter* 1(8):1 (July 1988).
10. Wurtman, J. J. *Managing Your Mind and Mood Through Food* (New York: Harper & Row, 1986).
11. "Hundreds Hope to Follow in Edison's Famous Footsteps," *Tri-City Herald* (January 27, 1985), p. D9.

12. "The Johari Window: A Model for Soliciting and Giving Feedback," in *The 1973 Annual Handbook for Group Facilitators*, J. E. Jones and J. W. Pfeiffer (San Diego, CA: University Associates, Inc., 1973), pp. 114–119.

Chapter 11 Breakthrough Environments: Setting the Climate

1. Amabile, T. M. *The Social Psychology of Creativity* (New York: Springer-Verlag, 1983).
2. Amabile, T. M., and S. S. Gryskiewicz. *Creativity in the R&D Laboratory* (Greensboro, NC: Center for Creative Leadership, 1987).
3. Gibb, J. R. *Trust: A New View of Personal and Organizational Development* (Los Angeles: The Guild of Tutors Press, 1978), p. 17.
4. Swiggett, R. L. "Structuring for innovation . . . and the Bottom Line," in *Structuring for Innovation: Creativity Week IX, 1986*, S. S. Gryskiewicz, B. Ghiselin, and M. W. Kiefaber, Eds. (Greensboro, NC: Center for Creative Leadership, 1987), pp. 61–74.
5. Hickman, C., and M. Silva. *Creating Excellence: Managing Corporate Culture, Strategy, and Change in the New Age* (New York: New American Library, 1984).
6. "Mimic Your Way to the Top," *Newsweek* (August 8, 1988), p. 50.
7. Kanter, R. M. *The Change Masters: Innovation and Entrepreneurship in the American Corporation* (New York: Simon & Schuster, 1983).
8. Kidder, T. *The Soul of a New Machine* (New York: Avon Books, 1981).
9. Mehrabian, A. *Public Places and Private Spaces: The Psychology of Work, Play, and Living Environments* (New York: Basic Books, 1976).
10. *Everybody's Business: A Fund of Retrievable Ideas for Humanizing Life in the Office* (Zeeland, MI: Herman Miller Research Corporation, 1985), pp. 65–76.
11. "Mothers' Enriched Environment Alters Brains of Unborn Rats," *Brain/Mind Bulletin* 12(7):1, 5 (March 1987).
12. *Everybody's Business: A Fund of Retrievable Ideas for Humanizing Life in the Office* (Zeeland, MI: Herman Miller Research Corporation, 1985), pp. 90–95.

Further Reading

On brain-mind studies:

- Fabian, J. P. "Creative Imagination Stimulation and Cognitive Style," Doctoral Dissertation, Washington State University, 1980.
- Wilber, K., Ed. *The Holographic Paradigm and Other Paradoxes: Exploring the Leading Edge of Science* (Boulder, CO: Shambhala Publications, Inc., 1982).

On brainstorming:

- Osborn, A. F. *Applied Imagination: Principles and Procedures of Creative Problem-Solving*, 3rd ed. (New York: Charles Scribner's Sons, 1963).

On celebrating creativity:

- Samples, B. *The Metaphoric Mind: A Celebration of Creative Consciousness* (Reading, MA: Addison-Wesley, 1976).

On the *Challenger*:

- Presidential Commission on the Space Shuttle *Challenger* Accident, W. P. Rogers, Chairman. *Report of the Presidential Commission on the Space Shuttle Challenger Accident* (Washington, DC: U.S. Government Printing Office, 1986).

On communication concept background:

- The Johari Window: A Model for Soliciting and Giving Feedback," in *The 1973 Annual Handbook for Group Facilitators*, J. E. Jones and J. W. Pfeiffer (San Diego, CA: University Associates, Inc., 1973), pp. 114–119.

On consciousness:

- Hampden-Turner, C. *Maps of the Mind* (New York: Collier Books, 1981).
- Samples, B. *Mind of Our Mother* (Reading, MA: Addison-Wesley, 1981).

On Data General's development of a computer:

- Kidder, T. *The Soul of a New Machine* (New York: Avon Books, 1981).

On developing people for creative thinking:

- Hickman, C., and M. Silva. *Creating Excellence: Managing Corporate Culture, Strategy, and Change in the New Age* (New York: New American Library, 1984).

On dreaming:

- Delaney, G. M. V. *Living Your Dreams* (San Francisco: Harper & Row, 1979).
- Evans, C., and P. Evans, Eds. *Landscapes of the Night: How and Why We Dream* (New York: The Viking Press, 1984).
- Garfield, P. *Creative Dreaming* (New York: Ballantine Books, 1972).

On Einstein:

- Koestler, A. *The Act of Creation* (New York: The Macmillan Company, 1964).
- Zukav, G. *The Dancing Wu Li Masters: An Overview of the New Physics* (New York: Morrow, 1979).

On engineers of the past:

- de Camp, L. S. *The Ancient Engineers* (New York: Ballantine Books, 1963).

On feeding the brain:

- Wurtman, J. J. *Managing Your Mind and Mood Through Food* (New York: Harper & Row, 1986).

On fostering a creative environment:

- Drucker, P. F. *Innovation and Entrepreneurship: Practice and Principles* (New York: Harper & Row, 1985).
- Hickman, C., and M. Silva. *Creating Excellence: Managing Corporate Culture, Strategy, and Change in the New Age* (New York: New American Library, 1984).

- Kanter, R. M. *The Change Masters: Innovation and Entrepreneurship in the American Corporation* (New York: Simon & Schuster, 1983).
- Peters, T. *Thriving on Chaos: Handbook for a Management Revolution* (New York: Alfred A. Knopf, 1988).
- Peters, T., and N. Austin. *A Passion for Excellence: The Leadership Difference* (New York: Warner, 1985).
- Peters, T. J., and R. H. Waterman. *In Search of Excellence: Lessons from America's Best-Run Companies* (New York: Harper & Row, 1982).

Of general interest:

- Adams, J. L. *Conceptual Blockbusting: A Guide to Better Ideas* (San Francisco: W. W. Freeman, 1974).
- Ghiselin, B. *The Creative Process: A Symposium* (Berkeley: University of California Press, 1982).
- Jewkes, J., D. Sawers, and R. Stillerman. *The Sources of Invention* (New York: W. W. Norton & Company, 1969).
- Kepner, C. H., and B. B. Tregoe. *The New Rational Manager* (Princeton: Kepner-Tregoe, Inc., 1981).

On imagery:

- Kosslyn S. M. "Aspects of a Cognitive Neuroscience of Mental Imagery," *Science* 240:1621–1616 (June 17, 1988).

On image making and meditation:

- Lazarus, A. *In the Mind's Eye: The Power of Imagery for Personal Enrichment* (New York: Guilford Press, 1984).
- Samuels, M., and N. Samuels. *Seeing with the Mind's Eye: The History, Techniques and Uses of Visualization* (New York: Random House, 1975).

On imaging the future:

- Fox, R., R. Lippitt, and E. Schindler-Rainman. *Towards a Humane Society: Images of Potentiality* (Fairfax, VA: NTL Learning Resources Corporation, 1973).

On Japanese KAISEN:

- Imai, M. *KAIZEN: The Key to Japan's Competitive Success* (New York: Random House, 1986).

- Ouchi, W. *Theory Z: How American Business Can Meet the Japanese Challenge* (Reading, MA: Addison-Wesley Publishing Company, 1981).

On media:

- Samples, B. *The Metaphoric Mind: A Celebration of Creative Consciousness* (Reading, MA: Addison-Wesley Publishing Company, 1976).

On mental games and exercises:

- Albrecht, K. *Brain Power: Learn to Improve Your Thinking Skills* (Englewood Cliffs, NJ: Prentice-Hall, 1980).
- Raudsepp, E. *Creative Growth Games* (New York: Perigee, 1977).
- Raudsepp, E. *More Creative Growth Games* (New York: Perigee, 1980).
- von Oech, R. *A Kick in the Seat of the Pants: Using Your Explorer, Artist, Judge, and Warrior to be More Creative* (New York: Harper & Row, 1986).
- von Oech, R. *A Whack on the Side of the Head: How to Unlock Your Mind for Innovation* (New York: Warner, 1983).

On mindmapping:

- Buzan, T. *Use Both Sides of Your Brain*, rev. ed. (New York: E. P. Dutton, 1983).
- Goldberg, P. *The Intuitive Edge: Understanding and Developing Intuition* (Los Angeles: J. P. Tarcher, 1983).
- Neimark, J. "Mindmapping: This Creative Way of Thinking Can Open Up a Whole New World of Opportunities," *Success* (June 1986), pp. 52–57.

On morphological boxes:

- Zwicky, F. *Discovery, Invention, Research: Through the Morphological Approach* (Toronto: The Macmillan Company, 1969).
- Zwicky, F. *Morphological Astronomy* (Berlin: Springer-Verlag, 1957).

On motivation:

- Amabile, T. M. *The Social Psychology of Creativity* (New York: Springer-Verlag, 1983).
- Amabile, T. M., and S. S. Gryskiewicz. *Creativity in the R&D Laboratory* (Greensboro, NC: Center for Creative Leadership, 1987).

FURTHER READING   281

On neurolinguistic programming:

- Grinder, J., and R. Bandler. *The Structure of Magic II* (Palo Alto, CA: Science and Behavior Books, Inc., 1976), pp. 3–26.

On open focus practice:

- Fritz, G., and L. Fehmi. *The Open Focus Handbook: The Self-Regulation of Attention in Biofeedback Training and Everyday Activities* (Princeton, NJ: Biofeedback Computers, Inc., 1982).
- *Open Focus: Cassette Training Series (Basic 1000)* (Princeton, NJ: Biofeedback Computers, 1977).

On order amid seeming chaos:

- Gleick, J. *Chaos: Making a New Science* (New York: Viking Penguin, Inc., 1987).

On other analysis methods:

- Altshuller, G. S. *Creativity As an Exact Science: The Theory of the Solution of Inventive Problems*, A. Williams, trans. (New York: Gordon & Breach, 1988).
- Keeney, R. L., and H. Raiffa. *Decisions with Multiple Objectives: Preferences and Value Tradeoffs* (New York: John Wiley & Sons, 1976).
- Kepner, C. H., and B. B. Tregoe. *The New Rational Manager* (Princeton: Kepner-Tregoe, Inc., 1981).

On other evaluational methods:

- Bailey, R. L. *Disciplined Creativity for Engineers* (Ann Arbor, MI: Ann Arbor Science Publishers, Inc., 1978).
- Keeney, R. L., and H. Raiffa. *Decisions with Multiple Objectives: Preferences and Value Tradeoffs* (New York: John Wiley & Sons, 1976).
- Kepner, C. H., and B. B. Tregoe. *The New Rational Manager* (Princeton: Kepner-Tregoe, Inc., 1981).
- Plunkett, L. C., and G. A. Hale. *The Proactive Manager: The Complete Book of Problem Solving and Decision Making* (New York: John Wiley & Sons, 1982).

On picture tour or confrontation:

- Geschka, H., U. Reibnitz, and K. Storik. *Idea Generation Methods: Creative Solutions to Business and Technical Problems* (Columbus, OH: Battelle Memorial Institute, 1981).
- McKim, R. H. *Thinking Visually* (Belmont, CA: Wadsworth, 1980).
- Samples, B., C. Charles, and D. Barnhart. *The WholeschoolBook: Teaching and Learning Late in the 20th Century* (Reading, MA: Addison-Wesley Publishing Company, 1977).

On psychological types:

- Keirsey, D., and M. Bates. *Please Understand Me: Character & Temperament Types* (Del Mar, CA: Prometheus Nemesis Book Company, 1984).

On the relationship between creativity and one's mental state:

- Arieti, S. *Creativity: The Magic Synthesis* (New York: Basic Books, 1976).
- "Mood Disorders Affect 80% of Well-Known Writers," *Brain/Mind Bulletin* 13(3):1 (December 1987).
- "The Ups and Downs of Creativity," *Time* (October 8, 1984).

On scientists and engineers embarking on the quest for discovery:

- Beveridge, W. I. B. *The Art of Scientific Investigation* (New York: Vintage Books, 1950).
- *The Illustrated Science and Invention Encyclopedia*, international ed. (Westport, CT: H. S. Stuttman Inc., 1983), 21 vols.
- Margenau, H., D. Bergamini, and Editors of Time-Life Books. *The Scientist* (New York: Time-Life Books, 1971).

On search keys generally:

- Beveridge, W. I. B. *The Art of Scientific Investigation* (New York: Vintage Books, 1950).
- Parnes, S. J., and H. F. Harding, Eds. *A Source Book for Creative Thinking* (New York: Charles Scribner's Sons, 1962).
- Perkins, D. N. *The Mind's Best Work* (Cambridge, MA: Harvard University Press, 1981).

On serendipity:

- "The Oops Factor," *Omni* 10(5):31 (February 1988).
- "Serendipity in Science: Survey Gathers Evidence," *Brain/Mind Bulletin* 11(17):1–3 (October 20, 1986).

On size of innovations:

- Gardner, J. W. *Self-Renewal: The Individual and the Innovative Society* (New York: W.W. Norton & Company, 1981).

On stimulating environments:

- Mehrabian, A. *Public Places and Private Spaces: The Psychology of Work, Play, and Living Environments* (New York: Basic Books, 1976).
- "Mothers' Enriched Environment Alters Brains of Unborn Rats," *Brain/Mind Bulletin* 12(7):1, 5 (March 1987).
- Ostrander, S., L. Schroeder, and N. Ostrander. *Superlearning* (New York: Dell Publishing Co., 1979).

On the story of insulin:

- Bliss, M. *The Discovery of Insulin* (Chicago: University of Chicago Press, 1982).

On superconductivity:

- "Feats of Inspiration and Originality" *Time* (October 26, 1987).
- "Not Doing It the Company Way," *Success* (January/February 1988).

On synectics:

- Gordon, W. J. J. *Synectics: The Development of Creative Capacity* (New York: Harper & Row, 1961).
- Prince, G. M. *The Practice of Creativity: A Manual for Dynamic Group Problem-Solving* (New York: Collier Books, 1970).

On toleration of nonsense:

- Zukav, G. *The Dancing Wu Li Masters: An Overview of the New Physics* (New York: Morrow, 1979).

On trust:

- Gibb, J. R. *Trust: A New View of Personal and Organizational Development* (Los Angeles: The Guild of Tutors Press, 1978).

On a view of intuition:

- Agor, W. H. *Intuitive Management: Integrating Left and Right Brain Management Skills* (Englewood Cliffs, NJ: Prentice-Hall, 1984).
- Goldberg, P. *The Intuitive Edge: Understanding and Developing Intuition* (Los Angeles: J. P. Tarcher, 1983).

On visual exercises:

- Edwards, B. *Drawing on the Right Side of the Brain* (Los Angeles: J. P. Tarcher, 1979).

References

Adams, J. L. *Conceptual Blockbusting: A Guide to Better Ideas* (San Francisco: W. W. Freeman, 1974).

"A New Look at Brain Machines: What the Experts Say," *The Omni Wholemind Newsletter* 1(6):1 (May 1988).

Agor, W. H. *Intuitive Management: Integrating Left and Right Brain Management Skills* (Englewood Cliffs, NJ: Prentice-Hall, 1984).

Albrecht, K. *Brain Power: Learn to Improve Your Thinking Skills* (Englewood Cliffs, NJ: Prentice-Hall, 1980).

Altshuller, G. S. *Creativity As an Exact Science: The Theory of the Solution of Inventive Problems*, A. Williams, trans. (New York: Gordon & Breach, 1988).

Amabile, T. M. *The Social Psychology of Creativity* (New York: Springer-Verlag, 1983).

Amabile, T. M., and S. S. Gryskiewicz. *Creativity in the R&D Laboratory* (Greensboro, NC: Center for Creative Leadership, 1987).

Arieti, S. *Creativity: The Magic Synthesis* (New York: Basic Books, 1976).

"Arts Enhance Scientific Intelligence, Study Says," *Brain/Mind Bulletin* 10(12):1–2 (July 8, 1985).

Bailey, R. L. *Disciplined Creativity for Engineers* (Ann Arbor, MI: Ann Arbor Science Publishers, Inc., 1978).

Beveridge, W. I. B. *The Art of Scientific Investigation* (New York: Vintage Books, 1950).

Bliss, M. *The Discovery of Insulin* (Chicago: University of Chicago Press, 1982).

"Breakthrough Ideas: How to Be an Entrepreneurial Einstein," *Success* (October 1987), pp. 46–53.

"Breeding New Ideas," *Newsweek* (August 8, 1988).

Brown, K. A. *Inventors at Work* (Redmond, WA: Tempus Books, 1988).

Bruner, J. S. *On Knowing: Essays for the Lefthand* (Cambridge, MA: Harvard University Press, 1963).

Buzan, T. *Use Both Sides of Your Brain*, rev. ed. (New York: E. P. Dutton, 1983).

Colligan, D. "Mindware Tools for Awareness," *The Omni Wholemind Newsletter* 1(8):1 (July 1988).

"'Confluent' Teaching Doubles Language Test Scores," *Brain/Mind Bulletin* 5(10):1,3 (April 7, 1980).

"Creative Thought Emerges from Emotional Theme," *Brain/Mind Bulletin* 7(7):2 (March 29, 1982).

de Bono, E. *Lateral Thinking: Creativity Step by Step* (New York: Harper & Row, 1970).

de Camp, L. S. *The Ancient Engineers* (New York: Ballantine Books, 1963).

Delaney, G. M. V. *Living Your Dreams* (San Francisco: Harper & Row, 1979).

"Dreams Led Barr to Pursue Melanin," *Brain/Mind Bulletin* 8(12/13):7 (July 11 and August 1, 1983).

Drucker, P. F. *Innovation and Entrepreneurship: Practice and Principles* (New York: Harper & Row, 1985).

Edwards, B. *Drawing on the Right Side of the Brain* (Los Angeles: J. P. Tarcher, 1979).

Evans, C., and P. Evans, Eds. *Landscapes of the Night: How and Why We Dream* (New York: The Viking Press, 1984).

Everybody's Business: A Fund of Retrievable Ideas for Humanizing Life in the Office (Zeeland, MI: Herman Miller Research Corporation, 1985).

Fabian, J. P. "Creative Imagination Stimulation and Cognitive Style," Doctoral Dissertation, Washington State University, 1980.

"Fast and Smart," *Time* (March 28, 1988).

"Feats of Inspiration and Originality" *Time* (October 26, 1987).

Fehmi, L., and G. Fritz. "Open Focus: The Attentional Foundation of Health and Well-Being," *Somatics* (Spring 1980), pp. 24–30.

Fox, R., R. Lippitt, and E. Schindler-Rainman. *Towards a Humane Society: Images of Potentiality* (Fairfax, VA: NTL Learning Resources Corporation, 1973).

Fox. *A Critique on Creativity in Science.*

Fritz, G., and L. Fehmi. *The Open Focus Handbook: The Self-Regulation of Attention in Biofeedback Training and Everyday Activities* (Princeton, NJ: Biofeedback Computers, Inc., 1982).

"Frog's Gift to Man," *U.S. News & World Report* (August 10, 1987).

Furnas, C. C., J. McCarthy, and Editors of Time-Life Books. *The Engineer* (New York: Time-Life Books, 1972).

Gardner, J. W. *Self-Renewal: The Individual and the Innovative Society* (New York: W.W. Norton & Company, 1981).

Garfield, P. *Creative Dreaming* (New York: Ballantine Books, 1974).

Geschka, H., U. Reibnitz, and K. Storik. *Idea Generation Methods: Creative Solutions to Business and Technical Problems* (Columbus, OH: Battelle Memorial Institute, 1981).

Ghiselin, B. *The Creative Process: A Symposium* (Berkeley: University of California Press, 1982).

Gibb, J. R. *Trust: A New View of Personal and Organizational Development* (Los Angeles: The Guild of Tutors Press, 1978).

Gleick, J. *Chaos: Making a New Science* (New York: Viking Penguin, Inc., 1987).

Goldberg, P. *The Intuitive Edge: Understanding and Developing Intuition* (Los Angeles: J. P. Tarcher, 1983).

Goodman, J., Ed. "The Fun Minute Manager," *Laughing Matters* 4(4):127–137.

Gordon, W. J. J. *Synectics: The Development of Creative Capacity* (New York: Harper & Row, 1961).

Grinder, J., and R. Bandler. *The Structure of Magic II* (Palo Alto, CA: Science and Behavior Books, Inc., 1976).

Hampden-Turner, C. *Maps of the Mind* (New York: Collier Books, 1981).

Hickman, C., and M. Silva. *Creating Excellence: Managing Corporate Culture, Strategy, and Change in the New Age* (New York: New American Library, 1984).

"Hundreds Hope to Follow in Edison's Famous Footsteps," *Tri-City Herald* (January 27, 1985), p. D9.

"The Idea Generator," *Success* (September 1988).

The Illustrated Science and Invention Encyclopedia, international ed. (Westport, CT: H. S. Stuttman Inc., 1983), 21 vols.

Imai, M. *KAIZEN: The Key to Japan's Competitive Success* (New York: Random House, 1986).

"Innovators: 100 Who've Made a Difference," *The Evergreen State Magazine* (Washington) 5(3):113–121 (November 1988).

Jewkes, J., D. Sawers, and R. Stillerman. *The Sources of Invention* (New York: W. W. Norton & Company, 1969).

"The Johari Window: A Model for Soliciting and Giving Feedback," in *The 1973 Annual Handbook for Group Facilitators,* J. E. Jones and J. W. Pfeiffer (San Diego, CA: University Associates, Inc., 1973), pp. 114–119.

Jung, C. G. *Man and His Symbols* (New York: Doubleday, 1964).

Kanter, R. M. *The Change Masters: Innovation and Entrepreneurship in the American Corporation* (New York: Simon & Schuster, 1983).

Keeney, R. L., and H. Raiffa. *Decisions with Multiple Objectives: Preferences and Value Tradeoffs* (New York: John Wiley & Sons, 1976).

Keirsey, D., and M. Bates. *Please Understand Me: Character & Temperament Types* (Del Mar, CA: Prometheus Nemesis Book Company, 1984).

Kepner, C. H., and B. B. Tregoe. *The New Rational Manager* (Princeton: Kepner-Tregoe, Inc., 1981).

Khatena, J., and E. P. Torrance. *Thinking Creatively with Sounds and Words: Norms-Technical Manual,* research ed. (Lexington, MA: Personnel Press, 1973).

Kidder, T. *The Soul of a New Machine* (New York: Avon Books, 1981).

Koestler, A. *The Act of Creation* (New York: The Macmillan Company, 1964).

Kosslyn, S. M. "Aspects of a Cognitive Neuroscience of Mental Imagery," *Science* 240:1621–1616 (June 17, 1988).

Lazarus, A. *In the Mind's Eye: The Power of Imagery for Personal Enrichment* (New York: Guilford Press, 1984).

Margenau, H., D. Bergamini, and Editors of Time-Life Books. *The Scientist* (New York: Time-Life Books, 1971).

Maslow, A. H. *Creativity in Self-Actualizing People.*

McKim, R. H. *Thinking Visually* (Belmont, CA: Wadsworth, 1980).

Mehrabian, A. *Public Places and Private Spaces: The Psychology of Work, Play, and Living Environments* (New York: Basic Books, 1976).

"Mimic Your Way to the Top," *Newsweek* (August 8, 1988), p. 50.

"Mood Disorders Affect 80% of Well-Known Writers," *Brain/Mind Bulletin* 13(3):1 (December 1987).

"Mothers' Enriched Environment Alters Brains of Unborn Rats," *Brain/Mind Bulletin* 12(7):1, 5 (March 1987).

Moyers, B. 1982 television creativity series.

Neimark, J. "Mindmapping: This Creative Way of Thinking Can Open Up a Whole New World of Opportunities," *Success* (June 1986), pp. 52–57.

"New Theory: Feelings Code, Organize Thinking," *Brain/Mind Bulletin* 7(6):1–2 (March 8, 1982).

"Nike: Designers, Researchers Encouraged to Pursue Their Ideas," *Oregonian* (March 12, 1989), p. L9.

"Not Doing It the Company Way," *Success* (January/February 1988).

O'Neill, J. J. *Prodigal Genius: The Life of Nikola Tesla* (New York: Ives Washburn, Inc., 1945).

"The Oops Factor," *Omni* 10(5):31 (February 1988).

Open Focus: Cassette Training Series (Basic 1000) (Princeton, NJ: Biofeedback Computers, 1977).

"'Optimal Performance' Similar in Athletics, Recovery," *Brain/Mind Bulletin* 5(8):1–2 (March 3, 1980).

Osborn, A. F. *Applied Imagination: Principles and Procedures of Creative Problem-Solving*, 3rd ed. (New York: Charles Scribner's Sons, 1963).

Ostrander, S., L. Schroeder, and N. Ostrander. *Superlearning* (New York: Dell Publishing Co., 1979).

Ouchi, W. *Theory Z: How American Business Can Meet the Japanese Challenge* (Reading, MA: Addison-Wesley Publishing Company, 1981).

Parnes, S. J., and H. F. Harding, Eds. *A Source Book for Creative Thinking* (New York: Charles Scribner's Sons, 1962).

Perkins, D. N. *The Mind's Best Work* (Cambridge, MA: Harvard University Press, 1981).

Peters, T. *Thriving on Chaos: Handbook for a Management Revolution* (New York: Alfred A. Knopf, 1988).

Peters, T., and N. Austin. *A Passion for Excellence: The Leadership Difference* (New York: Warner, 1985).

Peters, T. J., and R. H. Waterman. *In Search of Excellence: Lessons from America's Best-Run Companies* (New York: Harper & Row, 1982).

Plunkett, L. C., and G. A. Hale. *The Proactive Manager: The Complete Book of Problem Solving and Decision Making* (New York: John Wiley & Sons, 1982).

Poe, E. A. "Murders in the Rue Morgue," *Tales of Mystery and Imagination* (New York: The Mysterious Press, 1987), pp. 149–180.

Presidential Commission on the Space Shuttle *Challenger* Accident, W. P. Rogers, Chairman. *Report of the Presidential Commission on the Space Shuttle* Challenger *Accident* (Washington, DC: U.S. Government Printing Office, 1986).

Prince, G. M. *The Practice of Creativity: A Manual for Dynamic Group Problem-Solving* (New York: Collier Books, 1970).

Raudsepp, E. *Creative Growth Games* (New York: Perigee, 1977).

Raudsepp, E. *More Creative Growth Games* (New York: Perigee, 1980).

Ray, M., and R. Myers. *Creativity in Business* (Garden City, NY: Doubleday, 1986).

Rossman, J. *Industrial Creativity: The Psychology of the Inventor* (New Hyde Park, NY: University Books, 1964).

Samples, B. *The Metaphoric Mind: A Celebration of Creative Consciousness* (Reading, MA: Addison-Wesley Publishing Company, 1976).

Samples, B. *Mind of Our Mother* (Reading, MA: Addison-Wesley Publishing Company, 1981).

Samples, B., C. Charles, and D. Barnhart. *The Wholeschool Book: Teaching and Learning Late in the 20th Century* (Reading, MA: Addison-Wesley Publishing Company, 1977).

Samuels, M., and N. Samuels. *Seeing with the Mind's Eye: The History, Techniques and Uses of Visualization* (New York: Random House, 1975).

"Serendipity in Science: Survey Gathers Evidence," *Brain/Mind Bulletin* 11(17):1–3 (October 20, 1986).

"Special Issue: Prigogine's Science of Becoming," *Brain/Mind Bulletin* 4(13):1–4 (May 21, 1979).

"Special Issue: 'A New Science of Life,'" *Brain/Mind Bulletin* 6(13):1–4 (August 3, 1981).

Swiggett, R. L. "Structuring for innovation . . . and the Bottom Line," in *Structuring for Innovation: Creativity Week IX, 1986*, S. S. Gryskiewicz, B. Ghiselin, and M. W. Kiefaber, Eds. (Greensboro, NC: Center for Creative Leadership, 1987), pp. 61–74.

"The Ups and Downs of Creativity," *Time* (October 8, 1984).

von Oech, R. *A Kick in the Seat of the Pants: Using Your Explorer, Artist, Judge, and Warrior to be More Creative* (New York: Harper & Row, 1986).

von Oech, R. *A Whack on the Side of the Head: How to Unlock Your Mind for Innovation* (New York: Warner, 1983).

Whole Brain Conference, Key West, Florida, 1981.

Wilber, K., Ed. *The Holographic Paradigm and Other Paradoxes: Exploring the Leading Edge of Science* (Boulder, CO: Shambhala Publications, Inc., 1982).

Wurtman, J. J. *Managing Your Mind and Mood Through Food* (New York: Harper & Row, 1986).

Zukav, G. *The Dancing Wu Li Masters: An Overview of the New Physics* (New York: Morrow, 1979).

Zwicky, F. *Discovery, Invention, Research: Through the Morphological Approach* (Toronto: The Macmillan Company, 1969).

Zwicky, F. *Morphological Astronomy* (Berlin: Springer-Verlag, 1957).

Index

access 120–121
action phase
  attitude for approaching 212–214
  communication tips 214–218
  defined 206–207
  selling ideas 215, 217–218
  steps
    implement 211
    plan 207–211
aikido 225
algorithm 8–9
analogies 126–127, 153–161
analogy storm 153–161
analysis methods
  cause/consequence analysis 102–104
  force field analysis 99–101
  historical timeline 108
  observation 108–109
  other sources 111
  overview 94–96
  Pareto principle 104–107
  situation analysis 96–99
  value analysis 109–110
  see also analysis of situation in target phase
ancient inventions 3, 26, 76
arational and target phase 111–112
arational mode
  defined 39–40
  media for expressing 41–42
arational processes
  intuition 42–44
  preconscious 44–46
Archimedes 5
attention
  definition 28
  related to consciousness: *see* hard and soft focus; consciousness
auditory 48, 50–51
  defined 48
  exercise 51
  in thinking abilities 50

Bailey, Robert 16, 20, 65, 122
Banting, Frederick 213–214
Barr, Frank 45
Battista, Orlando 230
Bednorz, Georg 205–206, 212
Benjamin, R. B. 89
block breaker 59–60
blocks to creative thinking 60
  dull fantasies 65–66
  habitual thinking 57–59
  hard focus 55–56
  lack of emotion 67
  sensory blinders 108
  see mindsets
Bogen, Joseph 37
Bohm, David 37
brain, left hemisphere
  see brain-mind
brain, right hemisphere
  hemisphere studies 37–39
  modes, basic 37–39
  overview 31
  psychological types 32–36
  rational–arational modes contrasted 39–40
  receptive, beginner's mind 62
  see brain-mind
brainstorming 142–145
brainwriting 145–149
breakthrough qualities
  see childlikeness; creative elements; soft focus
breakthroughs
  creative elements 21–25
Buzan, Tony 149–150

calm 120
  incubate 134
Camplisson, Bill 126–127
Carnot, Sadi 56
case of parched land 91–94, 99–101, 103, 162–164, 193–199, 201–202, 210

cause/consequence analysis 102–104
Challenger disaster 97, 146–148, 178, 191
chaos 30–31
check phase
  assessing options 175–202
  communication tips 202–204
  consensus 188–189
  defined 176
  evaluation methods
    clarify 183–184
    method by screening 183
    sort, cull 184–188
    weigh 188–199
  evaluation screens 178–179
  getting started 176–180
  steps
    analyze potential problems 182
    clarify ideas 180–181
    search again 182
    sort, cull 181
    weigh, choose 181
  summary of evaluation methods 199–202
  using matrix 192–199
  who to involve 179–180
childlikeness
  creative child 60–61
  emotional coloring 66–67
  exploration 64–65
  fantasy 65–66
  playfulness 63–64
  receptive, beginner's mind 62
Christman, John 229
communication
  tips 112–116, 171–174, 202–204, 214–218
  visibility of ideas 46–48, 172–174, 215
consciousness 27–31
  defined 28
  inner world 29
  outer world 20–30
  planetary world 30–31
  see soft focus
consensus 188–189
Cray, Seymour 230

creative cycle
  for more imaginative ideas 72
  overview 23, 71–75
  see target; search; check; action phase
creative elements
  creative flow 21–25
  discovery processes 23–24, 71–75
  energy, creative power 21–23
  enhancements 24, 221–243
  environments 24, 247–270
  human instrument 21, 23, 26–53, 54–67
creative power 21–23
creative thinking
  choosing outcomes 19–21
  definitions 15–17
  disciplined 14–15
  myths 12–15
  who to involve 17–18
  see search phase
creativity
  defined 17
  role of genius 12–13
  role of mental illness 13–14
criteria
  arational approach 112
  as boundary conditions 86–87
  as *seen* by intuitives 86
  in evaluations methods 190–199

da Vinci, Leonardo 48, 63
Darwin, Charles 43
discovery
  application arenas 7–8
  experiencing 5
  getting started 6–10
  heuristic 9–10
  in past 3–4
  individual enhancements 223–234
  Japanese vs. American approach 10–12
  mental readiness 6–7
  pathways 8–10
  size of tasks 10–12
  see eureka 5–6

discovery cycle
    see creative cycle
discovery processes 23–24, 71–75
    creative cycle 23, 71–75
    creative strategies 24
    see search keys and phase
drawing
    see nonverbal, journal
dreams 44–46, 66, 120, 133
    Barr, Frank 45
    Hilprecht, Hermann 45
    Kekule 44

Edison, Thomas 5, 9, 48, 5–56
Eichelberger, Charles 235
Einstein, Albert 4, 51, 62, 63
emotions
    see childlikeness, mental illness
energy 21–23, 25
enhancements
    group
        facilitation 237–243
        see group genius and process
    individual
        add fuel 223
        aikido 225
        angle 229
        artistic ventures 227
        ask 229
        associate 223–224
        brain food 233–234
        collaborate 226
        computers 231
        creative space 225–226
        expand consciousness 224
        fantasize 65–66, 226
        idea garbage can 227
        join teams 229
        journal storm 226
        mediate 224–225, 232
        nonverbal 226, *see* nonverbal
        products, processes 232–233
        relax 224
        rummage 227
        self-talk 228–229
        serendipity 9–10, 211, 229–230

        tease brain 231–232
        train 227
        tunnel 230
        visualize 225, *see* visualization
environment 24
    building the organization 259–264
    creative space 225–226
    modeling 257–259
    overview 247–249
    philosophy and action 254–257
    stimulation 264–270
    trust and freedom 250–254
environmental skills 256–257
eureka 5–6
evaluation of options
    see check phase
Evans, Oliver 4, 212
exploration 64–65

facilitation 139, 143, 145, 150, 152, 183–199
    consensus 188–189
    in evaluation methods 183–199
    see communication; group genius and process; strategy formats in search phase
Faget, Maxime 127
fantasize 65–66, 226
Faraday, Michael 42–43, 47
Farnsworth, Philo 4, 58
Fehmi, Lester 55, 59, 163
Fleming, Alexander 108, 230
force field analysis 99–101, 182, 202, 209–210
force fit 158–159
Freberg, Stan 50
Fry, Art 267

Galileo 28
Gordon, William 153–154
Gray, William 38
Greatbatch, Wilson 129
group genius
    defined 235
    tap 234–243
group process

INDEX

group dynamics 238–240
group genius 234–243
  when to use 236–237
  meeting format 240–243
Gundlach, Bob 172–174

Hannan, Patrick 229
hard focus
  blocks to creative thinking 55–56
  defined 55
Harvey, William 126
Hatfield, Tinker 223
heuristic 9–10
  defined 17
Hilprecht, Hermann 45
historical timeline 108
human instrument
  mental equipment overview 26–27
humor 63–64, 117–118

idea generation
  see search phase
imagery
  see visualization
imagination
  see creative elements and thinking;
    search keys
  imagination stimulation 46
imaging the future 165–169
information processing
  see brain-mind; psychological types;
    arational processes; media for
    expressing
Ingham, Harry 238–239
innovation defined 17
intuition 42–44, 133–134
  defined 44
invention
  defined 17

journal 48, 133–134, 172–174, 226
judgment 100–101
Jung, Carl 30, 32–33

KAISEN 10–12
Kekule 44

Kettering, Charles 9
kinesthetic 51–52
  defined 51
  exercise 51
  in thinking abilities 50

Land, Edwin 4, 42
language
  defined 52
  exercise 52
Lippitt, Ron 168–169
Lorenz, Edward 30–31
Luft, Joseph 238–239

MacCready, Paul 64
matrix 129–131, 192–199
Maybeck, Wilhelm 4
media for expressing
  arational mode 41–42
  rational mode 41–42
  stimulating creativity 46–52, *see* also
    search keys; environments
    see auditory; kinesthetic; language; visual
meditation 224–225, 232
mental equipment
  see consciousness; brain-mind; arational processes; media for expressing
mental illness 13–14
mental processes 26–53
  consciousness 27–31
mind-brain
  see brain-mind
mindmapping 185, 149–153
mindsets 29, 57–59
  strategies for breaking 118–121
morphological box 131–133
Müller, Alex 205–206, 212
music 51, 233, 268–269
myths 12–15

neurolinguistic programming 46
nonverbal 112, 122, 125, 127–128, 145, 147–148, 150–151, 161–164, 226, 268–270

observation 108–109
outcomes 139–140
  see search: strategy formats
  zooming 80–82
Ovshinsky, Stanford 212
Owen, William 250

Pareto principle 104–107
Pasteur, Louis 229
Paxton, Floyd 13
picture tour 161–164
playfulness 63–64
Poe, Edgar Allen 78–80
power, creative 140–142, 169–170
preconscious 44–46, 133–134
  defined 44
Pribrim, Karl 28, 30, 37
Prigogine, Ilya 22–23, 30
Prince, George 153–154
problem–solving
  process
    see target phase
    situation-target interaction 77–80
    Situation-Target-Proposal 76–82
    zooming 80–82
  range of creative outcomes 19–21
  who to involve 17–18
psychological theories
  parent, adult, child 60–61
  psychological types 32–36
psychological types
  applied 148–149
  extroverts 33–34
  feelers 35
  introverts 34
  intuitives 34–35
  judgers 35–36
  perceivers 36
  regarding judgment 101
  selling ideas 215, 217–218
  sensors 34
  thinkers 35

questions
  question prompt 73, 89–91

question technique 114

Raphaelson, Sampson 15, 234–235
rational mode
  defined 39–40
  media for expressing 41–42
relaxation 38, 224
  relaxation and performance 38
Rico, Gabriele 149–150
Root–Bernstein 227
Rosen, Harold 215–216
Roy, Rustrum 229

search phase
  communication tips 171–174
  creative power
    higher 140–142, 169–170
    lower 140–142, 169–170
  keys
    access intuition, preconscious
      133–134
    analogies 126–127
    change form 127–128
    extract 129
    follow storming rules 123–124
    force fit 128
    incubate 134
    leap 128–129
    sparking storms 122–123
    stimulate, relax senses 125–126
    structure 129–133
    summary 134–135
    tinkering with angles 125
    see check phase steps
  steps
    choose, apply format 140–142
    determine outcome step 139–140
  strategies, overall
    access 120–121
    calm 120
    storm 119
  strategies, specific
    introduction 137–139
  strategy formats
    analogy storm 153–161

brainstorming 142–145
brainwriting 145–149
imaging the future 165–169
mindmapping 149–153
picture tour 161–164
summary 169–170
selling ideas 215, 217–218
serendipity 9–10, 211, 229–230
Sheldrake, Rupert 30
Silver, Spencer 267
situation defined 77
situation analysis 96–99
soft focus 54–55, 56–59, 224
  block breaker 59–60
  defined 56–57
  outrageous thinking 57–59
Sperry, Roger 37
stimulating creative thinking
  see search phase
storm
  defined 119
  rules 123–124
strategies
  see search phase
synergy 67, 114, 235
  see group genius

target phase
  arational approach 111–112
  communication tips 112–116
  defined 76, 77
  problem solving process 76–82
  steps
    analysis of situation 85, 91–93
    criteria 86–87, 93
    problem definition 83–85, 91
    question prompt 89–91, 94

    set target 87–88, 93
    solution availability 88–89, 93–94
  see group process; problem solving
teams
  composition 20
  effective teamwork case 150–151
  see group genius and process
  types 261–264
Tesla, Nikola 65
thinking
  creatively
    myths 12–15
    disciplined 14–15
  processes
    blocks 57–59
    outrageous thinking 57–59
  styles
    see media for expressing; psychological types

value analysis 109–110
visual
  defined 46
  exercise 47–48
  in thinking abilities 47
visualization 65–66, 225, 268–269
  visualization study 38
  see journal; nonverbal; visual

Wallace, Alfred 43
Wilson, Walter 231

Zasloff, Michael 9
zooming 80–82
Zwicky, Fritz 131–133